"MnO2 Composites for Supercapacitors"

M RASIQ

Title: "MnO2 Composites for Supercapacitors"

Author: M RASIQ

This book was printed and published by [Publisher's **M RASIQ**] in [2023]

ISBN: 9781805250746

ABSTRACT

The rapid penetration of portable consumer electronics and auto electrical devices in our society has led to a growing need for small-scale electrochemical storage (EES) devices to provide them with energy. The currently dominant energy sources are rechargeable Li ion batteries, which have a high energy density. They are powering almost all forms of consumer electronics and electric vehicles. The energy density of supercapacitors is considerably smaller, but supercapacitors are greatly appreciated for their high power density, long cycle life and safe operation. Supercapacitors are particularly useful in applications where a large amount of electrical energy needs to be stored or delivered quickly. Supercapacitor technology is developing quickly because of their increasing need in the ongoing electrification of our society.

Supercapacitors can be classified into two main classes, i.e., electrochemical double layer (EDL) capacitors and pseudocapacitors. The energy density of EDL capacitors is limited to the charge that can be stored in the so-called electrochemical double layer that is present in the electrolyte near the electrode surfaces. EDL capacitors typically employ metallic or graphitic electrodes. Pseudocapacitors also make use of fast and reversible faradaic reactions at the electrode surface. This requires the use of specific materials with a high concentration of surface redox sites. Since the EDL effect is also operative in pseudocapacitors, they can achieve significantly higher energy densities than EDL capacitors can. The desired pseudocapacitive materials are transition metal oxides. In particular, hydrous RuO_2, MnO_2, V_2O_5, several spinel phases and lamellar transition metal hydroxides can exhibit very high specific capacitances.

One of the most promising families of materials for pseudocapacitors is the Manganese oxide phases. Manganese oxide is abundantly present on Earth. It has a low toxicity and is an environmentally harmless element. It can crystallize in a number of polymorphs and morphologies. Manganese oxide is being used on large scale in Li ion batteries; however it also exhibits a very large pseudocapacitance. The electrochemical properties of MnO_2 based electrodes are determined by their crystal structure, morphology, conductivity, mass loading, and the type of electrolyte used. One of the major disadvantages of MnO_2 is its low electrical conductivity, which limits the pseudocapacitive redox reactions to the surface and near-surface layers. Careful engineering of the crystal structure, electrode morphology of MnO_2 based electrodes, composites with metals, metal oxides, conducting materials such as carbon, polymer are therefore of crucial importance to maximize their energy density and performance.

The primary motivation of the thesis is to prepare high performance supercapacitor electrodes through cost effective and environmental friendly methods. In this work, MnO_2 based nanocomposites with three different metal oxides such as, Nickel Oxide (NiO), Copper Oxide (CuO) and Iron Oxide (α-Fe_2O_3), have been synthesized via hydrothermal route. To improve their electrochemical performances, MnO_2 based metal oxides nanocomposites have been fabricated and investigated.

The Material characterization techniques includes X-ray diffraction (XRD), Fourier transform infrared (FT-IR) spectroscopy, high resolution scanning electron microscopy (HRSEM), energy dispersive X-ray (EDX) analysis, high resolution transmission electron microscopy (HRTEM), X-ray photoelectron spectroscopy (XPS), and Vibrating sample magnetometer (VSM) were utilized to study the properties of the bare nanoparticles and their nanocomposites. Electrochemical properties of bare nanoparticles and MnO_2 based nanocomposites were inspected by Cyclic Voltammetry (CV), Galvanostatic Charge and Discharge (GCD) and Electrochemical Impedance Spectroscopy (EIS). All the electrochemical experimentations were executed at room temperature. Cyclic voltammograms of the electrodes were recorded within a potential range between -0.1 to 0.4 V (vs. Ag/AgCl) in different scan rates from 5 to 320 mV/s. The specific capacitance of the electrodes was estimated over the charge/discharge curves as of the CP and CV measurements. The charge/discharge experiments were performed using chronopotentiometry, by applying different constant current range between -0.1 and 0.4 V (vs. Ag/AgCl). The electrochemical impedance spectroscopy was also studied in frequency region between 1Hz and 1MHz. The working electrode preparation consists of active material (80% by weight), activated carbon (10%) and Polyvinylidene Fluoride (PVDF) suspension (10%) ratio using the solvent of N-methyl-

2-pyrrolidone to make homogeneous slurry. Then the slurry from the mixture was pasted on the surface of nickel foil sheet current collector and dried out at 80 °C for 24h. All measurements were taken using an aqueous 1M KOH as electrolyte at ambient conditions with the electrode material prepared as a working electrode, a platinum wire as a counter electrode and Ag/AgCl as a reference electrode.

This thesis is arranged into five chapters. First chapter accentuates necessity to develop improved energy storage systems. In particular, a detailed discussion on the fundamental and classification of supercapacitors are presented. It also deals with the working principle, components and construction of supercapacitors. The most widely used electrode materials, electrolytes and separators are discussed. This chapter also provides comprehensive description of synthesis and characterization techniques for the preparation of MnO_2 based nanocomposites electrodes for the supercapacitor applications.

Second chapter is devoted to the fabrication of Manganese oxide/Nickel oxide (MnO_2/NiO) nanocomposites via hydrothermal route. The structural and morphological properties of the synthesized nanocomposites were investigated by various characterization techniques. Electrochemical studies indicate that the MnO_2/NiO nanocomposites exhibited improved electrochemical performance compared to the bare MnO_2 and NiO nanostructures. The enhanced performance was attributed the synergistic effect between the pseudocapacitance behaviour of MnO_2 and NiO nanoparticles.

In chapter three, the preparation of manganese oxide/copper oxide (MnO_2/CuO) nanocomposites was fabricated through hydrothermal method. The structural and morphological properties of the as-prepared nanocomposites were investigated using various characterization techniques such as XRD, HRSEM, EDX, TEM, FT-IR and XPS. The MnO_2/CuO composites exhibit improved electrochemical performances than the bare MnO_2 and CuO nanoparticles, which implies that the incorporation of CuO significantly enhance the electrochemical performances of the resultant MnO_2/CuO nanocomposites.

Fourth chapter deals with the synthesis and characterization of $MnO_2/\alpha\text{-}Fe_2O_3$ composites for supercapacitor applications. The preparation of nanocomposites involves hydrothermal heat treatment. The samples were characterized by XRD, FT-IR, HRSEM, EDX, HRTEM, XPS and VSM. The electrochemical performances of the $MnO_2/\alpha\text{-}Fe_2O_3$ composites were evaluated by CV, GCD and EIS techniques. The $MnO_2/\alpha\text{-}Fe_2O_3$ nanocomposites showed improved performance compared to bare MnO_2 and $\alpha\text{-}Fe_2O_3$ NPs. The improved electrochemical performance was attributed to the synergistic effects of pseudocapacitive behaviour between the MnO_2 and $\alpha\text{-}Fe_2O_3$ nanoparticles.

Chapter five summarizes the results obtained for three different metal oxide incorporated MnO_2 nanocomposites from various studies. In addition, this chapter suggests possible works that can be carried out, in future, to promote research on the developed MnO_2 based nanocomposites, understanding their importance in energy storage applications.

At the end of the thesis, relevant references and the papers published/presented in conferences are given. The papers published in peer reviewed internationals Journals are also attached.

SYMBOLS AND ABBREVIATIONS

SYMBOLS	ABBREVIATIONS
ε_r	Relative permittivity
D	Crystallite size
λ	Wavelength of X-ray
β	Full width of half maximum
θ	Angle
E_d	Energy density
P_d	Power density
ΔV	Potential window of discharge
I	Discharge current
nm	Nanometer
Kg	Kilogram
JCPDS	Joint Commission of Powder Diffraction Standards
Cs	Specific capacitance
V	Operating voltage
EC	Electrochemical Capacitor
EDLC	Electric Double Layer Capacitance
EES	Electrochemical Energy Storage
LIB	Lithium ion battery

NP	Nanoparticle
MnO_2	Manganese Dioxide
α-Fe_2O_3	Hematite or Iron Oxide
CuO	Copper Oxide
Na_2SO_4	Sodium sulfate
KOH	Potassium hydroxide
NMP	N-methyl-2-pyrrolidone
PVDF	Polyvinylidene Fluoride
CV	Cyclic Voltammetry
GCD	Galvanostatic Charge-Discharge
EIS	Electrochemical Impedance Spectroscopy
XRD	X-ray Diffraction
HRSEM	High Resolution Scanning Electron Microscopy
EDX	Energy Dispersive X-Ray Spectroscopy
FT-IR	Fourier Transform Infrared Spectroscopy
HRTEM	High Resolution Transmission Electron Microscopy
SAED	Selected Area Electron Diffraction
VSM	Vibrating Sample Magnetometer
XPS	X-Ray Photoelectron Spectroscopy

LIST OF TABLES

LIST OF FIGURES

TABLE OF CONTENTS

CHAPTER IV

SYNTHESIS AND CHARACTERIZATIONS OF $MnO_2/\alpha\text{-}Fe_2O_3$ NANOCOMPOSITE FOR SUPERCAPACITOR APPLICATION

CHAPTER V

SUMMARY AND SUGGESTION FOR FUTURE WORK

CHAPTER I

INTRODUCTION TO ELECTROCHEMICAL SUPERCAPACITORS

1.1 INTRODUCTION

Since the beginning of the 21st century, the rapid growth of population and the global economy has significantly increased the demand for energy consumption. Due to the serious impact of fossil fuels on the environment and the urgent need for energy storage electronics, it is necessary to develop efficient energy storage systems. Compared with traditional capacitors, supercapacitors (SCs) have relatively high energy densities and are lighter and smaller; compared with lithium-ion batteries, they have relatively high power densities due to their simple structure, charge and discharge speeds, and economic and environmental attractiveness, they are widely used in new energy vehicles, aerospace, defence science and technology, information technology, electronics industry, and many other areas [Zhang. L. L and Zhao. X. S 2009]. The comparison of the electrochemical behaviour for a typical supercapacitor and a typical battery are schematically given in Fig. 1.1 and detailed specifications are summarized in Table 1.1. The Ragone plots for different types of energy-storage devices are given in Fig. 1.2. Supercapacitors, also known as electrochemical capacitors, can complement or even replace batteries in some applications, because they can safely provide high power and rapid charging with extremely long cycle life (>100 000 cycles)[Simon. P and Gogotsi. Y 2008]. As a result, they have triggered a growing interest where fast charging, great cycling stability, and high power density is required. For example, supercapacitors are now used in heavy-duty vehicles, hybrid platforms for trucks and buses, load-levelling systems for intermittent renewable energy sources, and storing the regenerative braking energy of electric vehicles and light rail.

1.2 BRIEF HISTORIC OVERVIEW OF SUPERCAPACITORS

The earliest patent for supercapacitors dates back to 1957, when Becker [Becker. H. I 1957] prepared a high specific surface area of carbon-based supercapacitors. In 1990s, as people began to realize the potential of supercapacitor applications, it saw broad use as a hybrid power supply [Huggins. R. A 2000].

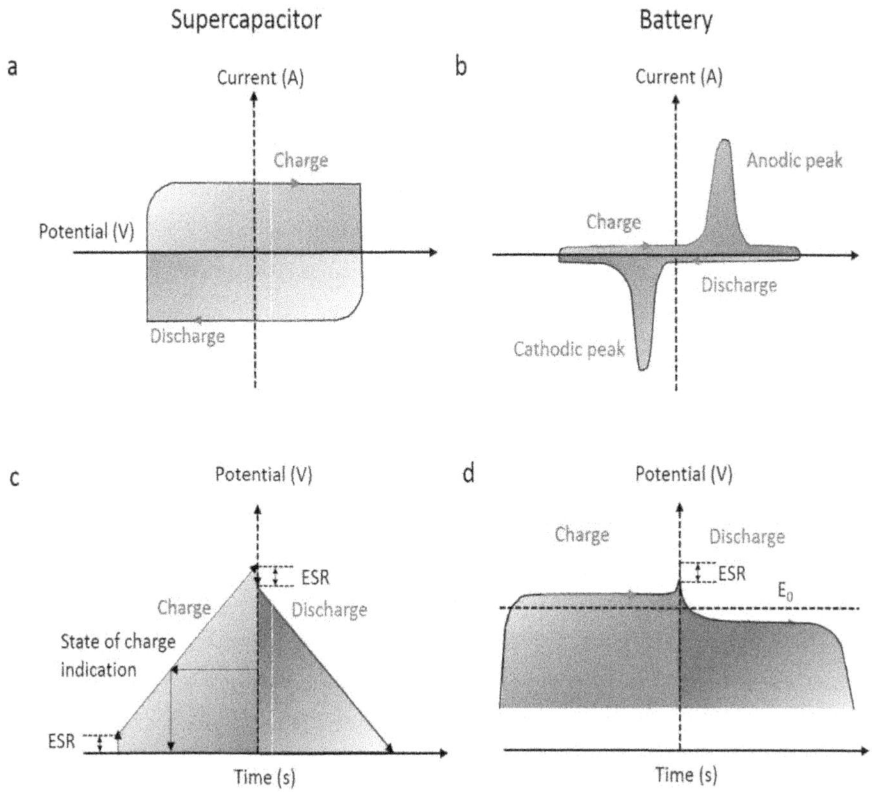

Fig.1.1: Comparison of the electrochemical behavior for a typical supercapacitor and a typical battery: (a and b) cyclic voltammogram curves and (c and d) galvanostatic charge–discharge curves.

Table 1.1: Comparison between supercapacitor, capacitor and battery

Parameters	Conventional Capacitor	Supercapacitor	Lithium-ion Battery
Energy storage	W-sec of energy	W-sec of energy	W-h of energy
Charge method	Voltage across terminals	Voltage across terminals	Constant current and constant voltage
Charge time	10^{-3}–10^{-6} s	1–10 s	10–60 min
Cell voltage	6–800 V	2.3 V–2.75 V/cell	1.2 V–4.2 V/cell
Specific energy (Wh/kg)	0.01 to 0.05	1 to 5	8 to 600
Charge/discharge efficiency	> 0.95	0.85–0.98	0.7–0.85
Specific power (W/kg)	< 100,000	Up to 10,000	1000–3000
Cycle life	> 500,000	1 million or 30,000 h	500 and higher
Operating temperature	− 20 to +100°C	− 40 to +85°C	− 20 to +65°C
Form factor	Small to large	Small	Small to large
Lifetime	> 100 k cycles	> 100 k cycles	150 to 1500 cycles
Weight	1 g to 10 kg	1 to 2 g	1 g to > 10 kg

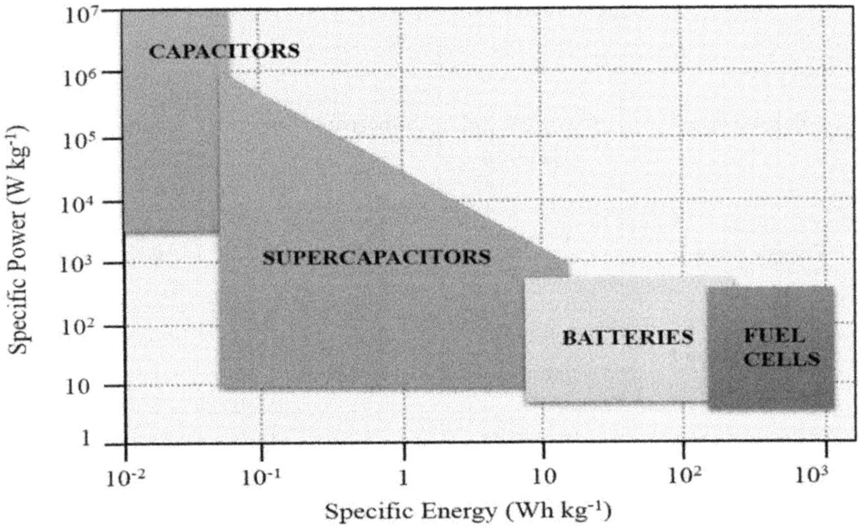

Fig. 1.2: Ragone plots for different types of energy-storage devices

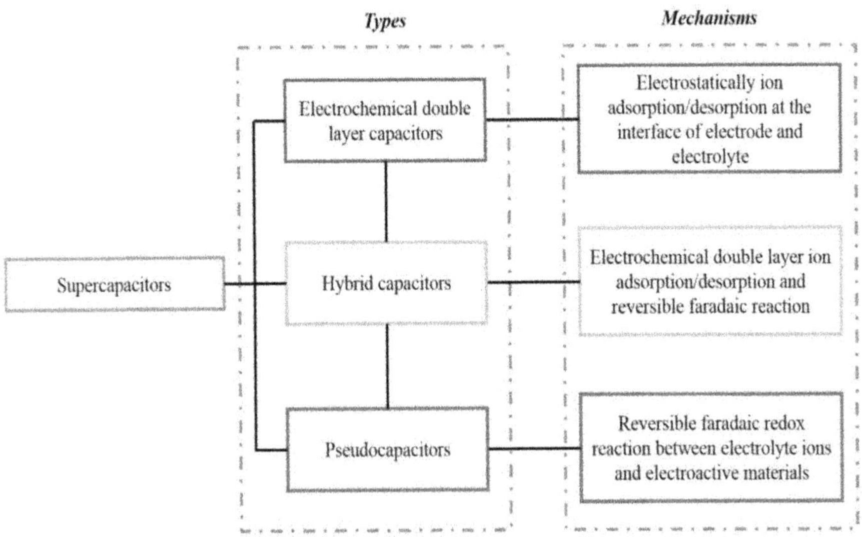

Fig. 1.3: Classifications of supercapacitors based on charge storage mechanism.

Now a days, supercapacitors are widely used in a variety of energy storage equipment; in the field of electrochemical energy storage devices, they have many more promising prospects and a broader application market. Today, Matsushita and NEC Corporation of Japan, as well as Maxwell Technologies of the United States, are major investors in the continued development of electrochemical capacitors [Wang. H et al. 2014]. Supercapacitors have huge application value and market potential and have become global hot spots for research.

1.3 TYPES OF SUPERCAPACITORS

Based on the charge storage mechanism involved, SCs are basically classified into two types; (a) Electric Double Layer Capacitors (EDLC) and (b) Pseudo-Capacitors (PC) or Faradaic Capacitors (FC) (c) Hybrid Capacitors. Fig. 1.3 shows the current classification of SCs based on charge storage mechanisms.

1.3.1. Electrical Double Layer Capacitor

Electrical double layer capacitor, which is made up of a carbon-based supercapacitor, derived its charge storage mechanism by electrostatic reaction where both positive and negative charges accumulate at the interfaces of electrode and electrolyte and form an electrical double layer (Fig. 1.4). The accumulation of the charges gives rise to the capacitance. In this charge storage mechanism, no transfer of electrons takes place and the capacitance depends on the surface area of the carbon-based materials. To obtain high specific capacitance, the large surface area of carbon-based materials such as carbon-based materials ranging from commercial activated carbons [Gamby. J et al. 2001], carbon aerogels [Wu. Z. S et al. 2012], templated carbon [Fuertes. A. B et al. 2005], to carbon nanomaterials, such as carbon nanotubes [Hu. S et al. 2012] and graphene [He. Y et al. 2013] are the most widely used and studied active materials for EDLCs owing to their high specific surface areas, desirable electrochemical stabilities, and open porosity to electrolyte ions. The specific capacitance of an EDLC is strongly dependent on the accessible surface area of the electrode materials and surface properties of the carbon materials. The capacitance of an EDLC electrode can be generally estimated according to eqn. 1.1.

$$\mathbf{C = \frac{\varepsilon_r \varepsilon_o}{d} A} \qquad \mathbf{(1.1)}$$

Where ε_r is the relative permittivity related to the liquid electrolyte used, ε_0 is the permittivity of vacuum, A is the effective surface area of the electrode materials which are accessible to the electrolyte ions, and d is the effective charge separation distance between the electrical double layers, i.e., the Debye length.

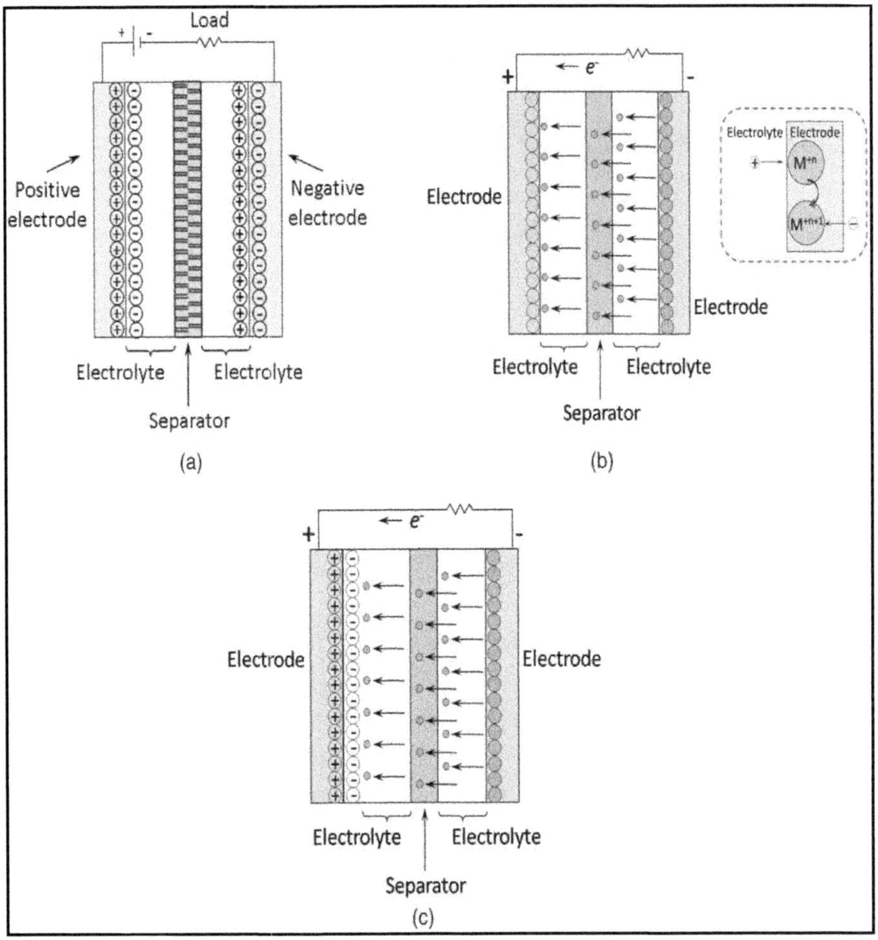

Fig.1.4: Schematic representation of charge-storage mechanisms for ELDC, Pseudocapacitors and Hybrid capacitors

1.3.2 Pseudocapacitors

On the other hand, pseudocapacitors, which make use of Conducting Polymers (CPs) and metal oxides as the materials for the electrodes, store the charge Faradaically. Pseudocapacitors are also called as redox supercapacitors that possess higher specific energy and can store more charges compared with EDLC as they can store charge via redox reaction in the bulk of the materials. However, pseudocapacitors have a lower specific power compared with EDLC as the redox reaction process takes longer time. For supercapacitors to be termed as high-performance supercapacitors, following properties should be owned: high specific energy and specific power, high specific capacitance, and long life cycle [Snook. GA et al. 2010].

It is mention here that although for several years nickel oxide/hydroxide and cobalt oxide/hydroxide have shown CV signature as that of battery, recently, active materials like MnO_2, NiO, Fe_3O_4 and RuO_2 have gained enormous attention for the supercapacitor application and are considered to be the true examples of pseudocapacitive materials. Among these oxides, RuO_2 and MnO_2 are the most widely studied active materials. Both the oxides are conductive in nature along with several oxidation states, and their charge storage mechanism is based on electro-adsorption of electrolyte cations or protons incorporation on their surface. The charge-storage mechanisms for pseudocapacitive electrodes are also shown in Fig. 1.4.

1.3.3 Hybrid Capacitors

Hybrid capacitor was mainly developed to reduce the drawbacks in both EDLC and pseudocapacitors. Need of hybrid capacitors is to exploit the relative advantages and alleviate the relative disadvantages of EDLC and pseudocapacitor to realize better performance characteristics. It utilizes both Faradaic and non-Faradaic processes to store charge. Hybrid capacitors have achieved energy and power densities greater than EDLCs without sacrificing the cycling stability and affordability that have limited the success of pseudocapacitor. Hybrid supercapacitors combine these two charge storage mechanisms (Faradaic and non-Faradaic), resulting in improved device characteristics. Generally, carbon based materials are good electrodes for EDLCs, while transition metal oxides and electrically conducting polymers are good candidates for developing Pseudocapacitor [Frackowiak. E 2007]. Now researchers focus on various types of hybrid capacitors, which are differed by their electrode configuration such as composite, asymmetric, and battery-type respectively. Recently, asymmetric or hybrid supercapacitor, development is the trend in electrochemical capacitors. Asymmetric supercapacitor can be fabricated with one electrode being of a double-layer carbon material and the other electrode being of a pseudo-capacitance material. It is possible to reach the high working voltage and high energy density by selecting a proper electrode material which, contributed to a significant increase in the overall energy density of the supercapacitor devices [Yang. P and Mai. W 2014]. The photographs of commercially available supercapacitors are given in Fig. 1.5.

1.4. SUPERCAPACITOR COMPONENTS

The desired performance of an EC supercapacitor ultimately begins with its components. These components can be tailored for specific applications; however, there are requirements for these components that are universal in the development of all EC systems. The aim of minimizing resistances while maximizing the potential to develop capacitive charge requires consideration of a number of factors for each component, including the electrode, electrolyte, separator, and current collector.

1.4.1 Electrolytes

Generally, supercapacitors use electrolytes that are usually classified into three types: (i) Aqueous, i.e., ions in water, (ii) Organic, salts in organic solvents, and (iii) Ionic liquid (IL), pure liquid salts [Zhong. C et al. 2015].

Electrolyte is one of the key parameters to determine the operating voltage of a supercapacitor. The rate capability and cycling performance also can be affected by the electrolyte ionic conductivity and electrochemical stability. Aqueous electrolytes can be classified into acids (e.g., H_2SO_4), alkali (e.g., NaOH), and neutral (e.g., Na_2SO_4). In addition, they usually exhibit the advantages of high ionic conductivity (up to 1 S cm^{-1}), low cost, ease of handling, and non-flammability. Due to their better electronic conductivity, higher dielectric constant, and larger accessible surface area of smaller aqueous electrolyte ions, supercapacitors with aqueous electrolytes can achieve higher capacitances than those using the same electrodes in non-aqueous electrolytes. However, there is still a restriction of voltage range due to the low thermodynamic stability window of water and moderate stability and durability during long-term cell operation. In addition, extra care needs to be taken for the selection of current collectors to avoid corrosion when choosing alkali or acidic electrolytes. While the low voltages of aqueous supercapacitors can only deliver limited capability to achieve high energy densities, they can do a very good job in the field of high power, low cost, excellent safety, and large-scale energy storage applications [Fic. K et al. 2012].

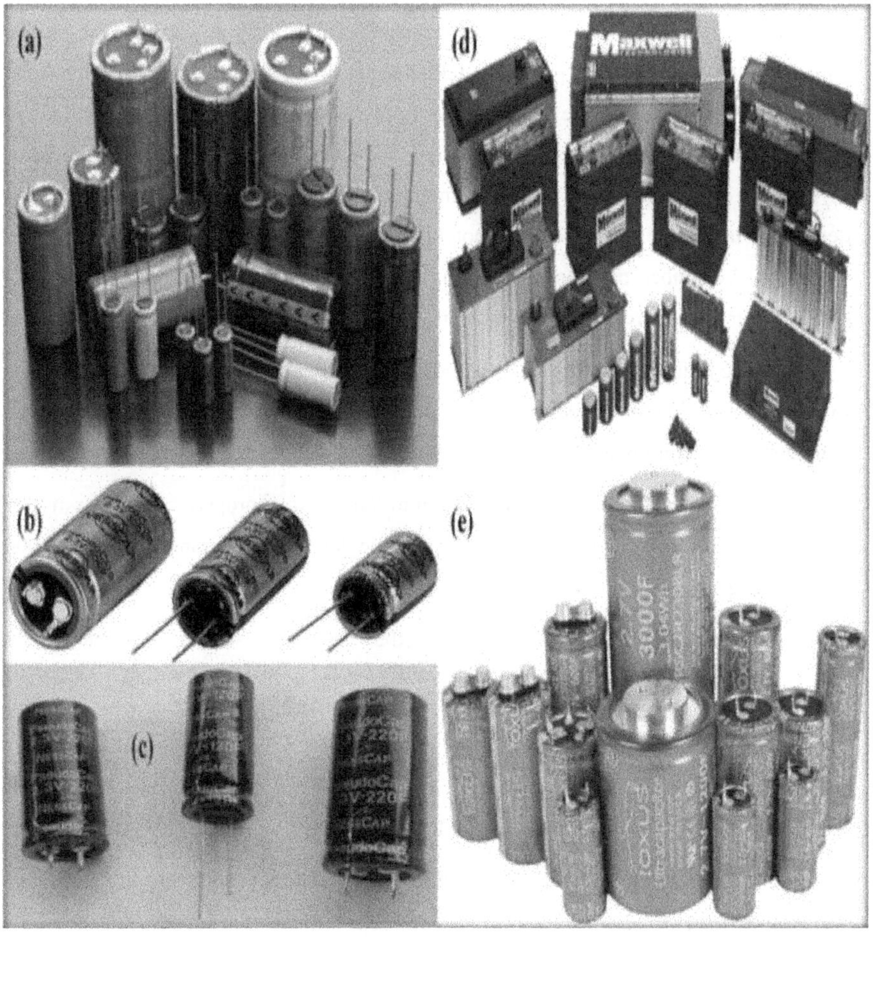

Fig. 1.5: Commercial supercapacitors as (a) EDLCs, (b) & (c) Pseudocapacitors, (d) & (e) Hybrid supercapacitors.

1.4.2 Electrode

An electrode for use in EC supercapacitors differs in a number of ways from those used in batteries. Primarily, current commercial supercapacitors are constructed with symmetric electrodes (i.e., both the anode and cathode are of the same material) where the ideal electric double-layer capacitance evolves as pure electrostatic charge with no oxidation–reduction processes taking place. Thus, distinguishing between the anode and cathode electrodes can simply apply which electrode is negatively and positively charged during discharge. This notation can change with the development of asymmetric capacitors; however, the distinction between anode and cathode is typically dependent upon the ideal operating potential window respective to the incorporated materials. A number of properties are required for a material to be of use as an electrode for EC supercapacitors, which are listed below [Kotz. R and Carlen. M 2000]:

The Cycling Capability (>105) and resulting highly stable cycle life; resistance to irreversible oxidation or reduction reactions along the surface of the electrode, which is important in avoiding loss of stability and performance

- In some cases, working in opposition to the previous property, the necessity for high SSA to develop the electric double layer, in the range of 1000–2000 $m^2 g^{-1}$.
- Electrochemical stability above the potential limits of electrolyte decomposition.
- Controlled pore size distribution to develop high capacitance while minimizing internal resistances.
- Surface wettability of the electrolyte.
- Mechanical reliability for its integration into a practical supercapacitor cell.
- Electrodes constructed using material types of either EDLC or pseudocapacitance have the respective benefits and challenges with their application for supercapacitors.

1.4.3 Separators

The choice of a thin, non-conducting separator must be considered when minimizing the electrolyte resistance contribution. Separators made of paper or polymer and ceramic or glasses are used with organic and aqueous electrolytes, respectively. Properties of a good separator include good transport of the electrolyte ions and sufficient mechanical resistance to the penetration of charged carbon particles traveling by electrophoresis [Zhi. M et al. 2013].

1.4.4 Current Collector

Current Collectors, typically metals and alloys with high electrical conductivity (e.g. Al, Fe, Cu, steel), are used for collection and efficient transport of current from the electrode to the applied load and, vice versa, recharging of the electrodes after use. Normally, a coating of current collector with the active electrode materials is done to enhance the contact between these two components [Wang. G et al. 2012].

1.5 ELECTROCHEMICAL MEASUREMENTS FOR SUPERCAPACITOR DEVICES

When characterizing the electrochemical performance of supercapacitors, one must be very careful to use the appropriate tests and evaluation methods. Generally, supercapacitor performance can be characterized by some essential parameters, such as specific capacitance (C), operating voltage (V), and equivalent series resistance (ESR), cycling stability, time constant, energy density, and power density. CV, GCD, and electrochemical impedance spectroscopy (EIS) are the three most commonly used electrochemical techniques to characterize the supercapacitor energy-storage performance.

1.5.1 Cyclic Voltammetry

CV provides a useful technique in screening potential capacitor materials in a relatively swift procedure. The technique is performed by cycling a linear potential sweep between two defined voltages determined by the electrolyte decomposition limits, and generally aided by a reference electrode in a three-electrode cell. The potential applied to the electrode varies linearly with time, yielding a time-dependent dynamic response current which is recorded by an oscilloscope or computer. The specific capacitance can be calculated by integration of the CV curves using eq 1.2. [Simon. P and Gogotsi. Y 2008]

$$\mathbf{C} = \frac{\int \mathbf{I}\,\mathbf{dV}}{\mathbf{v}\Delta\mathbf{V}} \qquad \mathbf{(1.2)}$$

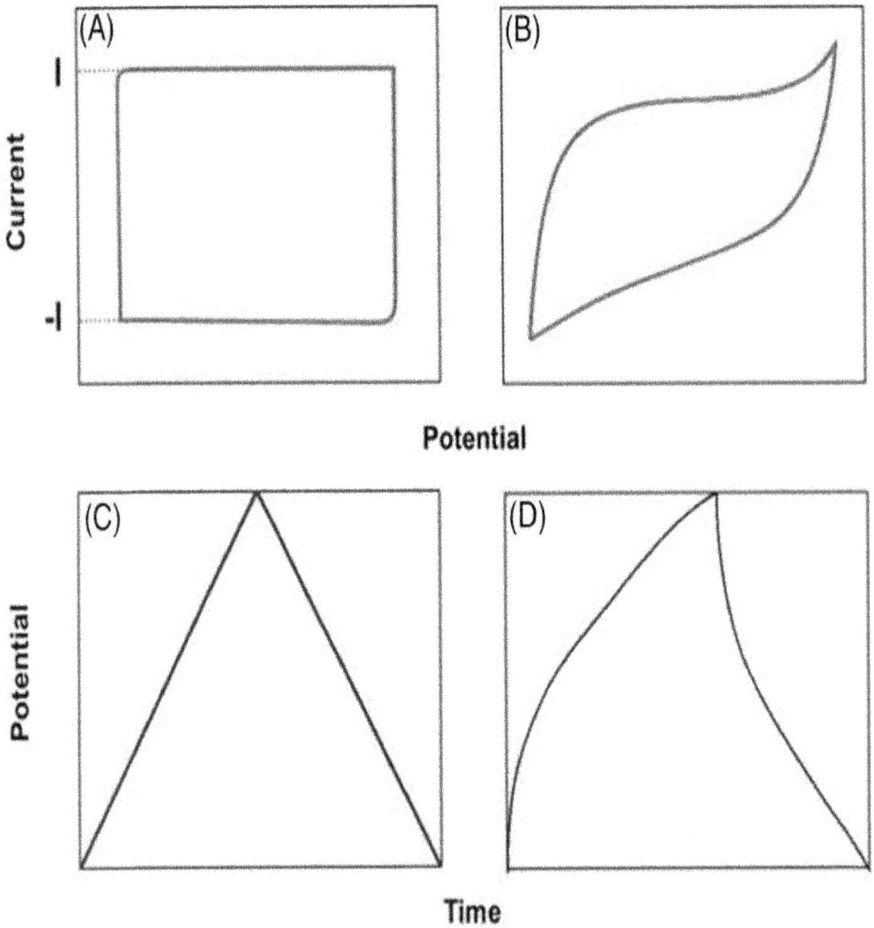

Fig. 1.6:Typical (A) cyclic voltammetry (CV) curves and (B) CV curve distortion due to Faradaic reactions (C) galvanostatic charge-discharge (GCD) curves for ideal supercapacitor; (D) GCD curve distortion due to Faradaic reactions.

Where I is the discharge current, i.e., the current below the X axis, v is the scan rate, and ΔV is the operating discharge potential range. In addition, the behaviour of the electrodes materials can also be illustrated from the shape of CV curves. Typically, ideal supercapacitor possesses perfect rectangular shape CV curves (Fig. 1.6A) with a constant current. However, the rectangular shape of CV curves can deviate due to the presence of Faradaic reaction (Fig. 1.6B) that originated from the CPs and metal oxides.

1.5.2 Galvanostatic Charge-Discharge

Interfacial charging of an electroactive material in contact with a solution electrolyte will behave capacitatively when a constant current density (i) is applied, developing a potential difference linearly with time across the cell. The working behaviour of GCD is more closely related to the electrochemical behaviour when an external load is typically applied to a supercapacitor for a majority of applications. Thus, the GCD test is the most commonly used and accurate method to determine the specific capacitance. It can be calculated by applying eq 1.3 [Simon. P and Gogotsi. Y 2008]

$$C = \frac{i \times \Delta t}{m \times \Delta V} \quad (1.3)$$

Where Δt and ΔV represent the discharging time (s) and potential window, respectively, whereas i represents the current applied (A) and m is the mass of the active materials on both electrodes (g). From the GCD curves, the amount of energy stored and the ability of a supercapacitor to be charged and discharged fast that are known as specific energy (Wh/kg) and specific power (W/kg), respectively, can be determined by using Equations 1.4 and 1.5. [Miller. J. R and Simon. P 2008]

$$E = \frac{1}{2} C_{SP} \Delta V^2 \quad (1.4)$$

$$P = \frac{E}{\Delta t} \quad (1.5)$$

Where Δt and ΔV represent the discharging time (s) and potential window, respectively. From the GCD curves, the reversibility of the electrode materials during process of charging/discharging can be deduced. The symmetrical GCD curves (Fig. 1.6C) indicate that the charging and discharging processes of the electrode materials are highly reversible. However, experimentally, the GCD curves (Fig. 1.6D) can be assigned due to the pseudocapactive behavior.

1.5.3 Electrochemical Impedance Spectroscopy

The examination of fundamental and applied characteristics of ECs can be performed using impedance spectroscopy over a wide range of frequencies (10 kHz to 10 MHz). ECs yield characteristic observations by plotting the complex plane of impedance (Nyquist). To determine the Csp of supercapacitor device from EIS, the graph of imaginary part of the impedance (−Z") with the reciprocal frequency (1/f) is plotted to determine the slope of the plot. The capacitance value is calculated from Equation [Ramya. R et al. 2012] followed by divided by the mass of the active materials of the electrodes to determine the capacitance per unit mass.

Apart from C_{sp}, EIS measurements can also provide information on the charge transfer resistance between the electrode/electrolyte interfaces, which can be obtained from the diameter of the semicircle at the high frequency of the Nyquist plot. At the low-frequency region of the Nyquist plot, the capacitive behaviour of the supercapacitor device can be deduced. Ideal capacitive behaviour is represented by the parallel line to the imaginary axis of the Nyquist plot [Wang. G et al. 2012].

Equivalent series resistance of the electrode materials which associated with the ionic resistance of electrolyte, resistance within electrode materials, and contact resistance between electrode materials and substrate can also be obtained from the Nyquist plot at intercept of semicircle on real axis. Apart from that, the system parameters of the electrochemical system can be deduced by fitting the EIS data with equivalent circuits. Fig. 1.7 shows the most common equivalent circuit model known as Randles circuit, which consists of resistance of solution (equivalent series resistance), resistance of charge transfer (R_{ct}), double layer capacitance (C_{dl}), and Warburg (W) [Wang. G et al. 2012].

1.6 KEY PARAMETERS TO ENHANCE THE SPECIFIC CAPACITANCE

The key parameters for high-efficiency supercapacitors mainly include specific capacitance, energy density, power density, rate capability and cycling stability [Yu. Z et al. 2015, Wang. G et al. 2012]. Therefore, the electrode materials for supercapacitors should possibly possess the following characteristics.

- **High specific surface area:** Charge absorption occurs at the interface between electrode and electrolyte, so the electrode with higher surface area effectively provides more active sites for ionic adsorption, contributing to an enhanced capacitance. Generally, materials with nanostructure can provide larger specific surface area than that with blocky structure.

- **Controllable porosity:** Reasonable pore size has access to the permeation of electrolyte and further boosts the utilization rate of electrode materials.
- **High electronic conductivity:** High conductivity could strengthen the responsiveness of electrode material to high scan rate or current density, and further to ensure an excellent rate capability.
- **Desirable electro active sites:** The more electroactive sites, the more ion adsorption, which could lead to high specific capacitance, for instance, nitrogen-doping, sulfur-doping, or co-doping.
- **High thermal and chemical stabilities:** The insolubility and structural stability of electrode materials in the cycle process are the preconditions for long cycle life

1.7 TRANSITION METAL OXIDES FOR SUPERCAPACITORS ELECTRODES

Transition metal oxides are unique class of solid functional materials in which oxygen are bound to the transition metals. The crystal structures of metal oxides constitute close-pack array structures in which oxygen anions are associated with metal cations occupying interstitial sites. Transition metal oxides usually consist of ionic bonding, show multiple oxidation states, and exhibit catalytic and semiconducting properties. With such properties and availability of metal cations, metal oxides actively participate in Faradaic redox reactions and thereby extensively used as electrode materials for supercapacitor application. Transition metals have a number of properties making them suitable for use as pseudocapacitors. Metal oxides, including ruthenium dioxide (RuO_2), iron oxide (Fe_2O_3), vanadium pentoxide (V_2O_5), tin dioxide (SnO_2), and manganese dioxide (MnO_2), among others, have been studied extensively for their development of pseudocapacitance. Some of the transition metal oxides with their theoretical specific capacitances are given in Fig.1.8

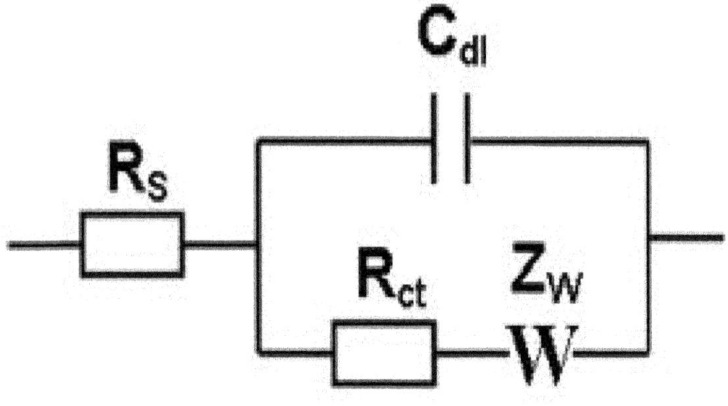

Fig. 1.7: Randles circuit

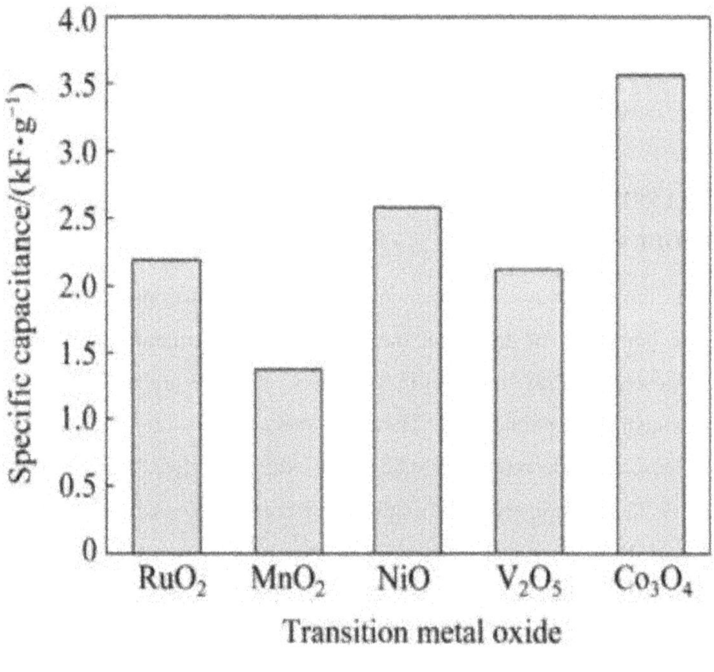

Fig. 1.8: Comparison of the theoretical specific capacitances of some transition metal oxides

1.7.1 RuO_2

Ruthenium oxide (RuO_2) is most widely studied metal oxide [Lin. Y.S et al. 2009] for EC material due to its various redox states, excellent electrical conductivity, a wide operating potential, considerably high capacitance, and an excellent cycle life. Both crystalline Ruthenium oxide (RuO_2) [Ramani. M et al. 2001] and amorphous hydrous Ruthenium oxide ($RuO_2{\cdot}xH_2O$) have been investigated to note a considerable drop in specific capacitance as a result of crystallization, despite the increase made to the intrinsic conductivity. A measurement of the specific capacitance of crystalline RuO_2, which is a d-band metallic conductor, gave 350 Fg^{-1} at $100mVs^{-1}$ [Lin. Y.S et al. 2009]. However, due to expensive nature the use of RuO_2 are limited in energy storage application. Presently, extensive scientific research is focused on materials alternative to RuO_2.

1.7.2 MnO_2

One of the most promising families of materials for pseudocapacitors is the manganese oxide phases [Conway. B.E et al. 1997]. It has a low toxicity and is an environmentally non harmful element. It can crystallize in a number of polymorphs and morphologies. Manganese oxide is being used on large scale in Li ion batteries; however it also exhibits a very large pseudocapacitance. Manganese dioxide is a stable manganese oxide, it has a theoretical specific capacitance of up to 1370 Fg^{-1} [Pandolfo. A.G and Hollenkamp. A.F 2006]. The electrochemical properties of MnO_2 based electrodes are determined by their crystal structure, morphology, conductivity, mass loading, and the type of electrolyte used.

1.7.3 Fe_2O_3

Currently, iron oxides work as the prospective anode materials in asymmetrical ECs due to the fact that compared with other metal oxides they present higher hydrogen evolution probability in aqueous solution. Fe_2O_3 being one of the most abundant material is significantly advantageous due to low cost, non-hazardous to living being, environmental friendliness and from safety point of view. However, their intrinsic poor conductivity and aggregated morphology restricts their extensive use as electrode material [Zhao. C et al. 2016].

1.7.4 NiO

Last decades, owing to the high conductivity pseudocapacitive performance and large surface area, nickel oxide has increasingly drawn our attention. Nickel oxide is highly abundant, low in toxicity, and exhibit desirable electrochemical properties for energy storage devices, including supercapacitor [Jiang. H et al. 2012]. NiO has the high theoretical capacity of 2573 Fg^{-1} mainly considered by high electrical conductivity.

1.7.5 CuO

Copper oxide (CuO) may also be considered as potential material for supercapacitor application, but the usage remain very limited as it suffers from major drawback of low electrical conductivity and unstable cycling performance which leads to relatively poor electrochemical properties. But coming to composite form CuO is the best candidate to induce an electrical conductivity in the nanocomposites. Recently, some researchers have been fabricated the electrodes by incorporation of copper hydroxide or copper oxide into carbonaceous materials, metals and conducting polymers [Zhu. S et al. 2015].

1.7.6 Co_3O_4

Co_3O_4 is considered to be another promising material for supercapacitor application and thereby extensive studies have been carried out on this material. It is noteworthy to mention that the environmentally benign and low cost nature along with high theoretical specific capacitance of 3560 Fg^{-1} make the material much desirable for supercapacitor [Ujjain. SK et al. 2015]. Table 1.2 shows the different values obtained from different metal oxide based supercapacitors.

1.8 CRYSTAL STRUCTURE OF MnO_2

MnO_2 is a class of supercapacitor materials, and the MnO_2 crystal structure has a direct relationship with its electrochemical performance. MnO_2 has a variety of crystalline phases, in which the MnO_2 [MnO_6] octahedron is the basic unit: O atoms located on each of the six corners, and Mn atoms in the center [Feng. Q et al. 1999]. These chains make different crystal forms in different bonding arrangements via corner sharing. The common crystal phases of MnO_2 are α, β, γ, δ, and λ, and the size of the pore structure of each manganese oxide with distinct different crystal phases is different. The common crystal structure of MnO_2 is shown in Fig. 1.9

Table 1.2: Metal oxide-based electrode materials with electrochemical performances.

Electrodes	Electrolyte	Maximum Specific Capacitance (F/g)	Current density (A/g)	Scanning rate (mV/s)	Retention with Cycle	Ref.
RuO_2	1 M H_2SO_4	734	10	1	91% @200	[Jang. J.H et al. 2006]
MnO_2	0.1 M Na_2SO_4	698	-	50	73.6% @1500	[Pang. S.C et al. 2000]
$Ca_3Co_2O_6$	-	563	-	5		[Ali. G.A.M et al. 2015]
SnO_2	0.1 M Na_2SO_4	285	0.3	10	88% @1000	[Rajendra Prasad. K et al. 2004]
In_2O_3	1 M Na_2SO_3	190	3	10	@1000	[Prasad. K.R et al. 2004]
Bi_2O_3	1 M NaOH	98	-	100	@1000	[Gujar. T.P et al. 2006]
V_2O_5	2 M KCl	350	-	-	@100	[Lee. H.Y and Goodenough. J.B 1999]
Fe_3O_4	1 M Na_2SO_3	170	-	2	-	[Wang. S.Y et al. 2006]
	1 M Na_2SO_4	25	-		-	

1.9 MnO_2 BASED NANOCOMPOSITES FOR SUPERCAPACITORS

Nanostructured MnO_2 electrode materials exhibit the unique properties of nanomaterials, and are a focus of the research because of their high specific capacitance, abundant raw materials, and environmental friendliness. However, their conductivity is generally low, which seriously affects their specific capacity, cycle life, and scalability. To improve the conductivity and cycling stability of MnO_2, it is advantageous to prepare composite materials in order to make use of synergistic effects. Common MnO_2 composite materials contain metals, other transition metal oxides, conductive polymer and carbon nanocomposite components are the most widely used [Zhou. H et al. 2015]. In recent years, many researchers have reported the charge storage of MnO_2-based composite materials for supercapacitors in order to summarize the latest developments of MnO_2 to design high performance supercapacitor nanocomposite electrodes for the preparation and assembly of MnO_2-based supercapacitor electrode.

1.9.1 MnO_2-Carbon Composites for SCs

Forming MnO_2-based hybrid structures with highly conductive carbons is one of the most adopted methods to improve the capacitor performance of MnO_2 electrodes whose poor conductivities limit their capacitance, cycling life and rate performance. Carbon materials (carbon nanotubes, graphene, porous carbon, carbon nanofibers, carbon spheres, carbon nanofoam/aerogel) are promising candidates to couple with MnO_2 nanostructures. Various carbon materials have been investigated as the supporting material for the deposition of MnO_2 nanostructures due to their outstanding properties, such as high electrical conductivity, chemical stability in both acidic and alkaline medium, and light weight. These features enable carbon materials to be the most popular scaffold materials for fabrication of MnO_2 electrodes.

These carbons can act both as high surface area mechanical support and current collector for the active MnO_2 phase to improve the conductivity of the electrode. It is noted that the carbon scaffolds can also act as active materials for charge storage, simultaneously increasing the whole capacitance of the composite electrodes [Zhou. H et al. 2015].

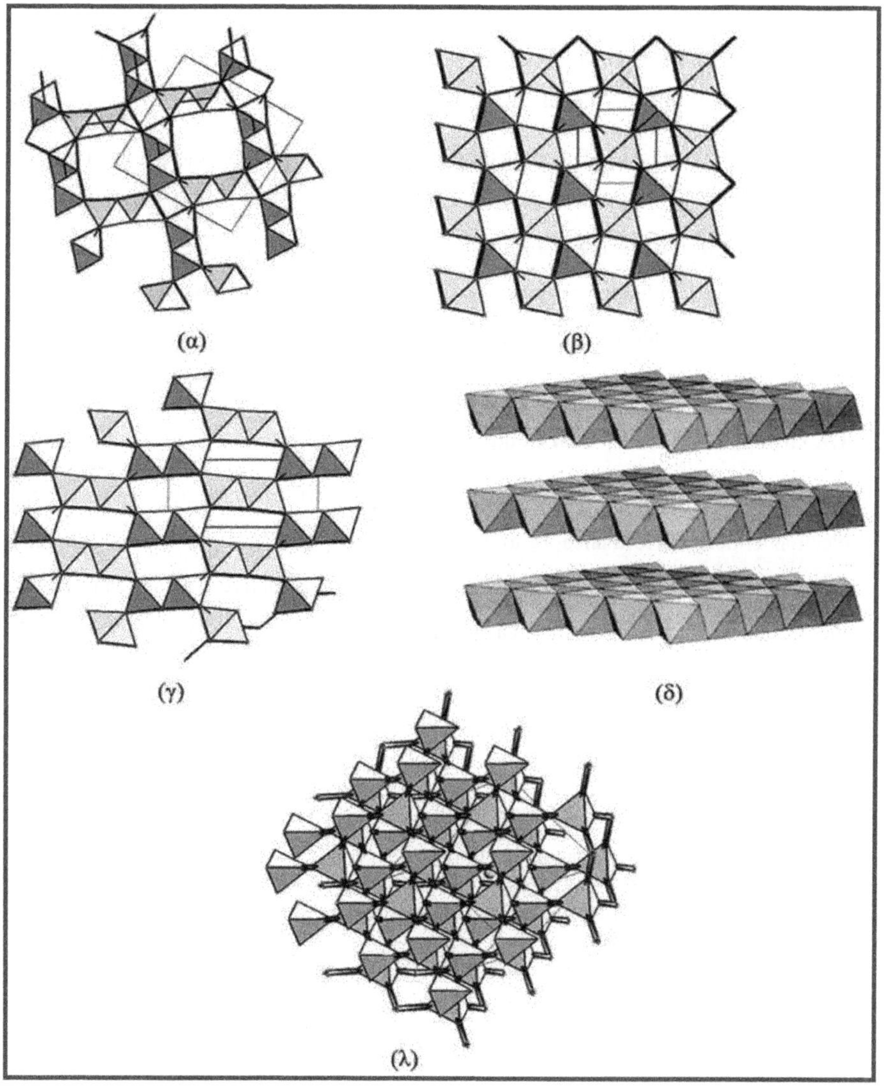

Fig. 1.9: Crystal structures of $\alpha, \beta, \gamma, \delta$ and λ- MnO_2.

1.9.2 MnO_2-Conducting Polymer Composite for SCs

Conducting polymers, such as polyaniline (PANI), Poly (3, 4-thylenedioxythiophene) (PEDOT), Polypyrrole (PPy), are also considered as a category of promising supporting materials to couple with MnO_2 for advanced electrochemical energy storage [Moussa. M et al. 2017].

The advantages of conducting polymers are mainly including the good conductivity, controllable synthesis, feasibility of forming uniform and porous structures, and the pseudo-capacitive characters. Based on these properties, the combination of conducting polymers with MnO_2 would harvest the synergetic effects between polymers and oxides, thus providing some superiority for the supercapacitor application.

1.9.3 MnO_2-Condutive Metal Composite for SCs

Another effective way to improve electrical conductivity and to boost electrochemical performance is to conformably deposit MnO_2 nanostructures onto highly conductive and well-ordered metal micro-/nanostructures, which can serve as the mechanical support with greatly enhanced electron and ionic transport. The transition metals (Al, Cu and Ni, stainless steel, Mn film) are also good scaffolds to form MnO_2-based hybrid films or composites by conformal coating or penetration for SCs [Simon. P and Gogotsi. Y 2008]. The challenge of the application of porous metal for preparing the MnO_2-conducting metal composite electrode is its high production cost, which limits its practical applications.

1.9.4 MnO_2-Metal Oxide/Hydroxide Composite for SCs

Incorporation of certain metal oxide/hydroxide with MnO_2 to fabricate MnO_2-based composite is one of the effective ways to improve the electrochemical property. Through optimizing preparation parameters, generated synergistic effects may enlarge specific surface area, facilitate electrons and ions transport, add active sites or improve cycling stability. Up to now, various MnO_2-metal oxide/hydroxide composite with different geometrical attributes and morphological forms have been employed as electrodes for SCs. Combining MnO_2 with many other transition oxides (SnO_2, CuO,

NiO, Fe_2O_3) are also developed by designing novel composite pseudo-capacitive systems [Simon. P and Gogotsi. Y 2008, Zhou. H et al. 2015].

Fabrication of MnO_2-metal oxide/hydroxide composite electrode with high specific surface area and porous configuration has been intensively studied, and a series of good results has been achieved. As described above, many metal oxides or hydroxides (Co_3O_4, NiO, $Ni(OH)_2$, TiO_2, ZnO, etc.) have higher electrical conductivity than MnO_2, but with narrower operating potential window. Therefore, by combining these different metal oxide/hydroxides with MnO_2, synergic properties are expected. Research has shown that the combination of two types of materials not only provides the composite electrode material with stronger redox ability, but also enhances the stability. Most importantly, the higher conductivity of the other metal oxide/hydroxides can serve as the backbone to support and provide effective electrical connection to MnO_2, which can produce a synergistic effect of all individual constituents. According to these reasons, tuning the microstructure, crystallinity, and conductivity of the MnO_2-based electrodes by constructing binary or ternary composites could be an efficient way to improve the performance of MnO_2-based supercapacitors.

1.10 PREPARATION METHODS FOR METAL OXIDE ELECTRODE MATERIALS

Nano sized metal oxide electrode materials are widely studied. Much research has been carried out on the preparation methods of metal oxide nanostructured electrode materials. At present, the some important preparation methods are the hydrothermal/solvent thermal method, the redox method, the sol–gel method, the microwave-assisted method, and the electrodeposition method.

1.10.1 Sol–Gel Method

The sol–gel (solution−gelation) method is highly versatile and one of the most popular methods for the synthesis of oxide materials with high homogeneity and reactivity. The sol–gel process involves four main steps: (i) the formation of a sol; (ii) transforming a sol to a gel; (iii) converting a gel into xerogel or aerogel; and (iv) transforming the xerogel/aerogel into the desired particles.

1.10.2 Precipitation and Co-precipitation Method

Precipitation or co-precipitation is a frequently used, especially in the industry, and technically very important method to prepare the oxide particles of fine size and homogeneity. Precipitation is often used to synthesize oxides with sample structure, for instance aluminium oxides, silicon oxides, titanium oxides, zinc oxides, and indium oxides. The major advantages of this method lie on (i) the possibility of creating very pure materials and (ii) the flexibility of the process regarding the final product quality. Meanwhile, co-precipitation is more suitable for the preparation of more complex oxides with two or more cations by a definite stoichiometry [Gujar. T.P et al. 2006, Lee. H.Y and Goodenough. J.B 1999, Wang. S.Y et al. 2006]. Both precipitation and co-precipitation have a similar process, which involves the following steps: (i) choosing inorganic salts with high solubility (such as chlorides, carbonates, and nitrates) as metal precursors; (ii) reaction with precipitating agents (sodium bicarbonate, ammonia, etc.) to obtain the precipitation of hydroxides; and (iii) separating the precipitation from the solution followed by thermal treatments.

1.10.3 Hydrothermal Method

Hydrothermal synthesis is a widely used technology to crystallize ceramic materials directly from aqueous solution by utilizing single or heterogeneous phase reactions and elaborate control of thermodynamic parameters (e.g. Pressure, Temperature, and Composition). The process of hydrothermal synthesis includes the following steps: (i) Preparation of the precursor by mixing the soluble metal salt in aqueous solution; (ii) transferring the mixed solution in to an autoclave and reacting under elevated temperature (usually 100–300°C) and pressure (>100 kPa); and (iii) completing the reactions followed by filtration, washing, and drying. Hydrothermal synthesis possesses a host of merits over other oxide-preparation methods. All forms of oxides can be synthesized by this method, namely powders, single crystals, coatings, and nanofiber. Besides, the morphologies (sizes and shapes) of oxides can also be precisely controlled. Hydrothermal synthesis is also a time-saving and energy-efficient method owing to the unnecessity of some time- and energy-consuming steps (such ashigh-temperature calcination and milling) often used in other methods.

1.10.4 Microwave Sintering

Microwave sintering is a method that converts electromagnetic energy to thermal energy. Energy is delivered directly to the material through molecular interaction with the electromagnetic field; hence, the energy efficiency is very high. Besides, the bulk ceramic would own more uniformed refined microstructures, and there will be fewer thermal mismatch stresses and the firing temperatures would be lower than solid-state sintering.

1.10.5 Electrospinning Method

Electrospinning method is a versatile technique to produce nanofibers with controllable diameter, morphology and surface topology from a variety of organic and inorganic materials. It gives the impression of being a very simple and, therefore, easily controlled technique for the production of fibers with dimensions down to the nanometer range. First, polymers will be surveyed as fiber-forming materials. Later, materials such as metals, ceramics, and glasses will be considered as fiber precursors. Electro spinning mainly makes use of the electrostatic repulsions between surface charges to reduce the diameter of a visco elastic jet or a glassy filament. A number of functions can be incorporated into the nanofibers by electro spinning to expand their potential applications with the only prerequisite that the second phase can be soluble or well dispersed in the initial solution.

1.10.6 Electrodeposition Method

The electrodeposition method proceeds in the solution containing the targeting ions which will be deposited on the anodic or cathodic electrode substrates by controlling the deposition rate, and thus the target material films with different density and morphologies will be obtained. The electrodeposition technique has been widely used because of its simplicity, the elimination of polymer binders, non-polluting nature and modest reaction conditions. The advantages and disadvantages of various synthesis methods are summarized in Table 1.3.

Table 1.3: Advantages and disadvantages of various synthesis methods

Fabrication techniques	Merits	Demerits
Hydrothermal/ Solvothermal	➢ Provides highly monodispersed particles with controllable size and morphology ➢ Large-scale production possible ➢ Produces high purity materials ➢ No toxic chemicals involved	➢ Require high temperature and pressure
Chemical bath deposition	➢ Large-scale production is possible ➢ Growth rate is faster than hydrothermal and solvothermal	➢ Not applicable to all metal oxides
Sol-gel	➢ Large-scale production is possible ➢ Produces high purity materials ➢ Low processing temperature	➢ Formed dense film without using template, so challenging in producing porous films
Chemical precipitation	➢ Large-scale production is possible	➢ Difficult to control the morphology
Electrochemical deposition	➢ Uniform morphology can be obtained	➢ Large-scale production is not possible
Electrophoretic deposition	➢ Produces high purity materials ➢ Applicable to wide range of materials (metals, ceramics and polymers)	➢ Water electrolysis from aqueous suspensions
Microwave-assisted	➢ Large-scale production is possible ➢ Reduces the reaction time	➢ Difficult to control the phase and morphology

1.11 APPLICATION OF SUPERCAPACITORS

The ever-improving performance of supercapacitor leads to constant increment in their applications. Last decade has seen the extensive dominance of rechargeable batteries in the energy storage business. The growth in demand for enhanced energy storage for diverse applications regarding electronic portability to hybrid electric vehicles is rapidly increasing. The need in the sector of energy storage is performance optimization and cost reduction. Due to unique storage capability of supercapacitors, it has been widely used in various applications like electric drives, Computers and memory backup chips, radar system, electromagnetic pulse weapons, automotive sector, UPS, traction, electric vehicles, SSD's, LED flashlights etc. Herein, two examples of EC application in an electric bus (e-bus) are briefly introduced to clarify the future application trend of ECs.

- The fast-charge e-bus is one important direction for public city traffic systems. ECs are the most suitable candidate for such applications, as they can be charged within seconds. Aowei Technology Co., Ltd (Shanghai, China) has successfully applied $Ni(OH)_2$/AC hybrid ECs (or hybrid supercapacitors) in a trolley bus (Fig. 1.10). This type of trolley bus can be fully charged within a very short time of 90 s and drive 7.9 km at a time with an average and maximum speed of 22 and 44.8 km h^{-1}, respectively. It takes 16.5 s to increase the speed from 0 to 40 km h^{-1}.
- Very recently, EDLCs have been applied in an electric tramcar by the Chinese CSR Co., Ltd (Fig. 1.11). This type of electric tramcar can be charged in a very short time of 30 seconds and drive 3–5 kilometres at a time. Without the chemical reactions of a high-capacity supercapacitor, there is no explosion or fire risk, the cycle number can reach 1 million, and its life can be up to 10 years. The characteristics of energy conservation and environmental protection for this type of electric tramcar can alleviate urban traffic pressure and reduce the environmental pollution.

Fig. 1.10: Trolley bus based on $Ni(OH)_2$//AC supercapacitor

Fig. 1.11: Electric tramcar based on AC/AC supercapacitor

1.12. CONCLUSION

This chapter briefly discussed the introduction of electrochemical supercapacitors from more technical and practical aspects and the crystallization of their development in the last decades. The fundamental electrochemical theory, specific capacitance calculations from electrochemical results, basic components involved in the electrochemical systems is discussed. MnO_2 based nanocomposites have been widely studied to optimize the material cost, easy process with simultaneous improvement in the electrochemical performances. This chapter is also comprehensively discussed the recent progresses on fabrication and designing, electrochemical performances and achievements of MnO_2 and its nanocomposite as electrode materials for supercapacitors will be beneficial for further research designing high performance supercapacitor devices.

CHAPTER II

SYNTHESIS AND CHARACTERIZATION OF MnO_2/NiO NANOCOMPOSITES FOR SUPERCAPACITOR APPLICATION

2.1. INTRODUCTION

With the rapid development of the global economy, the depletion of non-renewable energy like coal and oil is resulting in more and more serious environmental crisis. Therefore, it is urgent to explore renewable, clean energy sources such as wind and solar energy and their intermittent nature evokes a high demand for effective and reliable energy storage technologies. In this energy-dependent world, electrochemical devices for energy storage play an energetic role in overcoming the depleting fossil fuels [Augustyn. V et al. 2014]. Supercapacitors received great interest in academic and industrial world in recent years owing to their fast charge/discharge rate, high power density and extensive cycling life [Lukatskaya. M. R et al. 2016, Liu. Y et al. 2014, Wang. G. P et al. 2012]. Supercapacitors and batteries are similar arrangements with negative and positive electrodes and separators soaked in electrolytes [Wu. H. B et al. 2016, Yu. A et al. 2013]. Supercapacitors are broadly categorized into two types, i.e, double layer electrical capacitors (EDLC) and pseudocapacitors [Nooshabadi. M.S and Zahedi. F 2017]. In the case of EDLC, the capacitance comes from accumulation of charges on electrode-electrode interfaces. Furthermore, the storage of pseudocapacitance energy is carried out by transferring the faradic charges between electrolyte and electrode, due to reversible multielectronic redox phase reactions [Deng. L et al. 2013]. Consequently, the energy density of pseudocapacitors is remarkably higher in the category of capacitor technologies. Electrical double-layer capacitor electrode materials include various carbon materials such as activated carbon, carbon nanotubes (CNT), and graphene. Some redox-active metal oxides (MOs) (RuO_2, NiO, Fe_2O_3, Co_3O_4, and MnO_2) and conducting polymers (polyaniline, polypyrrole, and polythiophene-based conducting polymers) have been used as electrode materials for pseudocapacitors. Among them, MOs have attracted great attention due to easy preparation, low cost, high chemical stability and thermal stability.

Varieties of synthetic techniques have been introduced to synthesize metal oxide nanostructure, such as hydrothermal, sol-gel, sonochemical, microwave-assisted and electrodeposition techniques. Among them, hydrothermal method is most widely applied for the fine fabrication of nanostructured metal oxides in term of its highly dispersible, size-controllable, effective and high-yield properties.

Among the metal oxide electrode materials for pseudocapacitors, MnO_2 has been widely studied due to low cost, low toxicity, large theoretical capacity (~ 1370 Fg^{-1}) and natural abundance [Zhao. L et al. 2014]. For enhancement in specific capacity of MnO_2, synthesis of nanostructured electrodes based on MnO_2 with new morphologies, porous hierarchical structures, high surface area and large pore volume were explored using various methods [Zhao. L et al. 2014, Tang. C.L et al. 2015, Zhu. S. J et al. 2015]. MnO_2-based nanomaterials with granulometry, morphology, crystallinity, high specific surface area and good controllable electrical conductivity are critical to improving capacity, rate capability [Hu. Y and Wang. J 2015, Li. W et al. 2014, Lu. X et al. 2011]. It is clear that advances in a pseudocapacitor technology should be made to take advantage of those benefits towards the large-scale applications. To improve the MnO_2-based electrodes, fabrication of MnO_2-based compounds incorporating with porous metal substrates, conductive polymers, carbon and graphene etc. [Xiao. K et al. 2014, Wang. C et al. 2014, Chen. L. F et al. 2013]. Nickel oxide (NiO) is widely studied as one of the most promising pseudocapacitive materials possessing an unusually high specific theoretical capacity (2573 Fg^{-1}) [Liu et al. 2012] and an excellent reactivity of electrochemical reaction. The morphologically controlled preparation of NiO nanostructures of three distinct morphologies (nanoflower, nanosphere, nanoparticle) were synthesized using sol-gel technique and exploited their morphology dependant supercapacitor performance. The flower-like NiO with high surface area showed the best supercapacitor behavior by delivering 480 Fg^{-1} specific capacitance at 0.5 Ag^{-1} current density [Kim. S.I et al. 2013].

In recent times, much attention has been increased to enhance the specific capacitance of the supercapacitors, exploring new materials with the rational intention of the multi-component combination [Zhao. X et al. 2012, Liu. J.P et al. 2011], which provides synergistic effect of all the individual component, as well as the control 3D hierarchical heterostructures, which maintain a large external surface and provide effectual and fast routes for the transport of ions and electrons, thus high capacitance at fast charging– discharging rates. Although the individual capacitive properties of NiO and MnO_2 have been widely demonstrated [Kang. J et al. 2013, Yuan. C.Z et al. 2009, Qian. Y et al. 2014], the electrochemical properties of the compounds has also received some interest and is expected to be a desirable material for the high performance supercapacitor. For example, Yi. T-F et al (2020) reported the 3D spherical NiO@MnO_2 composites with core-shell structure prepared by a hydrothermal route; the sample exhibits a high specific capacitance of 1219.2 Fg^{-1} at 20 Ag^{-1} and excellent stability with a capacitance retention rate of 76.7% after 10,000 cycles. Bi. Y et al. (2018) developed the nanostructured NiO/ MnO_2@ graphite composite synthesized via an ultrafast and energy-efficient microwave solid-state approach. The NiO/MnO_2@ graphite showed excellent cycling stability with capacitance retention of 140% (82.22 mAh g^{-1}) of the initial value after 1600 cycles as compared to 16.56 mAh g^{-1} (28.3%) after 400 cycles for NiO@graphite. Moreover, the specific capacitance of NiO/MnO_2@graphite increased from 58.47 to 102.78 mAh g^{-1} (maximum value) after an activation period. The excellent electrochemical performances of NiO/MnO_2@graphite composite can be attributed due to the presence of MnO_2 that have the stability over long cycle life. Liu et al. (2009) prepared nanostructured MnO_2-NiO composite by the sol–gel and chemical deposition method. They found that introducing a proper proportion of the NiO into MnO_2 to form an electroactive material is an effective method to achieve high performance electrochemical capacitor. Liu. J et al. (2012) reported the synthesis of MnO_2–NiO nanoflake-assembled tubular array on stainless steel substrate to function as pseudocapacitor electrode by programmed three-dimensional (3D) interfacial reactions, in which the ZnO nanowire array was employed as the low-cost sacrificial template. Jiang et al. (2012) presented a novel hybrid structure of porous MnO_2–$Ni(OH)_2$ composite with a nanoflake surface by a green, surfactant-free and scalable strategy. The optimal MnO_2–$Ni(OH)_2$ composite exhibited a very high specific capacitance with excellent rate and cycling stability in 1 M Na_2SO_4 aqueous solution (355.5 Fg^{-1}), and even in 1 M KOH aqueous solution (487.4 Fg^{-1}). Liu et al. (2012) reported the fabrication of NiO/MnO_2 hybrid nanostructures in the stainless steel substrate by interfacing reactions to pseudocapacitors, which have a high area capacity of 0.35 Fcm^{-2}. Li et al. (2015) established fabrication of hierarchical matrices of NiO/MnO_2 nanosheets on carbon fibre paper through the hydrothermal process, which could provide a real specific capacity of 1.0 Fcm^{-2} (494.8 Fg^{-1}). Junjiao Chen et al. (2016) fabricated $NiO@MnO_2$ core/shell nanocomposites by a two-step method which exhibited a specific capacitance of 266.7 Fg^{-1} at 0.5 Ag^{-1} and excellent cycling stability (81.7% retention after 2000 cycles at 1 Ag^{-1}). Yuhong Bi et al. (2018) reported microwave synthesis of NiO/MnO_2 composite which revealed a specific capacitance of 102.78 mAh g^{-1} at 1 Ag^{-1} and superior cycling stability with specific capacitance retention of 140% over more than 1500 cycles. Shuang Xi et al. (2017) synthesized NiO/MnO_2 Core-Shell Nanoflakes on Carbon Cloth with areal capacitance of 316.37 mF/cm^2 and special gravimetric capacitance of 204.3 F/g at the scan rate of 50 mV/s. The electrode retains 89% of its initial discharge capacitance after 2200 cycles with >97% Coulombic efficiency. However, the synthetic procedures for these MnO_2 NiO core-shell structures are relatively complicated. Therefore, a simple wet chemical approach to prepare MnO_2/NiO nanocomposite is essential. However, the synthetic procedures for these NiO/MnO_2 core-shell structures are relatively complicated. Therefore, a simple wet chemical approach to prepare MnO_2/NiO nanocomposite is essential.

The high energy density (energy stored per unit mass), extended power density (rapid energy delivery and quick charging), good rate capability (capacitance retention over fast charging) and excellent cycle stability (material stability or structural stability of electrode material) are the major factors which scale the performance and reliability of the device over other charge storage devices. Hence achieving high power delivery for a prolonged time with extended cyclic stability or reproducibility over aging is highly demanded to meet the current energy crisis. Here, a promising electrode material has been fabricated through compositing MnO_2 and NiO by facile hydrothermal route. The structure, morphology, surface state and magnetic properties of bare MnO_2, NiO and MnO_2/NiO nanocomposite were examined by XRD, FT-IR, HRSEM, EDX, TEM, XPS and VSM techniques. The electrochemical properties were investigated by CV, GCD and EIS techniques. The results emphasize that the MnO_2/NiO nanocomposites are potential electrode materials for supercapacitors.

2.2. EXPERIMENTAL

2.2.1 Preparation of MnO_2 Nanoparticles

The MnO_2 nanoparticles were synthesized by a facile hydrothermal method. Typically, $MnCl_2 \cdot 4H_2O$ and urea were dissolved in 50 mL of deionized water to form a homogeneous solution. Then, the above solution was transferred into Teflon autoclave and hydrothermally heated at 120°C for 24 hours. The resulting precipitate powders were washed three times with 100 mL of DI water and three times in ethanol. Finally, the powder was dried at 80° C for 12h. Finally, the sample was obtained by annealing the dried product at 400°C for 3 h.

2.2.2 Preparation of NiO Nanoparticles

For the synthesis of NiO nanoparticles, 3.239g $NiCl_2 \cdot 6H_2O$ was dissolved in 50 mL distilled water. Then, 1.502 g of urea was added into the obtained Nickel chloride solution under magnetic stirring for 30 min (Nickel Chloride to Urea ratio= 1:1). Then the mixed solution was transferred into a Teflon-lined autoclave (100mL) sealed and kept at 120°C for 24 h. Centrifugation is employed to separate the precipitate of the mixed solution with deionized water and ethanol. Finally, the powder was dried at 80°C for 24 hours. Resultant NiO nanopowder was obtained by annealing the product at 400°C for 2 h.

2.2.3 Preparation of MnO_2/NiO Nanocomposite

For the MnO_2/NiO nanocomposite preparation, the above synthesized NiO was dissolved in 20 mL distilled water by magnetic stirring, then $MnCl_2 \cdot 4H_2O$ and urea was mixed into the above solution. After continuously stirring for 1h, the mixed solution was transferred to a 150 mL capable autoclave and heated at 120 °C for 24 h. The black colour precipitate obtained from above process was collected by filtration and washing with deionized water and ethanol. The resulting final product was dried at 80°C for 12 h and annealed at 400 °C for 4h. The detailed synthesis procedure is also given as flowchart in Fig. 2.1.

2.2.4 Characterizations

The X-ray diffraction (XRD) analysis of the prepared samples were performed on Rigaku D/max-kA diffractometer using Cu-Kα radiation (λ = 1.54060 Å) over a 2θ range of 20° to 80° with the 0.05° step size in order to investigate the structure of MnO_2, NiO nanoparticles and MnO_2/NiO nanocomposites. The chemical functional group of the samples were recorded by Fourier transform infrared spectroscopy (FTIR) on a Perkin Elmer 1600 instrument in the range of 400 to 4000 cm^{-1}. The surface morphology and elemental analysis of the sample were performed by high resolution scanning electron microscope (HRSEM, FEI Quanta 200) which is equipped with EDX and the TEM images were obtained on a Philips CM200FEG field emission microscope. The surface electronic states presence was analysed by X-ray photoelectron spectroscopy (XPS, Perkin-Elmer PHI 5000 Versa Probe III). The room temperature magnetization measurements were performed using a vibrating sample magnetometer (VSM, Lake Shore VSM-7410) with a maximum applied magnetic field of 15 kOe.

2.2.5 Electrochemical Measurements and Electrode Preparation

Electrochemical properties of prepared samples were inspected by CH-Instrument Model 6008E electrochemical work station to study the cyclic voltammetry (CV), Galvanostatic Charge and Discharge (GCD) test and Electrochemical Impedance Spectroscopy (EIS) results. All the electrochemical experiments were executed at room temperature. Cyclic voltammograms of the electrodes were recorded within a potential range between -0.1 to 0.4 V (vs. Ag/AgCl) in different scan rates from 5 to 320mV/s. The specific capacitances of the electrodes were calculated over the charge/discharge curves as of the CP and CV measurements. The charge/discharge experiments were performed using chronopotentiometry, by applying different constant current range between -0.1 and 0.4 V (vs. Ag/AgCl). The electrochemical impedance was also studied in frequency region between 1Hz and 1MHz.

The working electrode preparation consists of active material (80% by weight), activated carbon (10%) and Polyvinylidene Fluoride (PVDF) suspension (10%) ratio using the solvent of N-methyl-2-pyrrolidone to make homogeneous slurry. Then the slurry from the mixture was pasted on the surface of nickel foil sheet current collector and dried out at 80 °C for 24h. All measurements were measured using an aqueous 1M KOH as electrolyte at ambient conditions with the electrode material prepared as a working electrode, a platinum wire as a counter electrode and Ag/AgCl as a reference electrode.

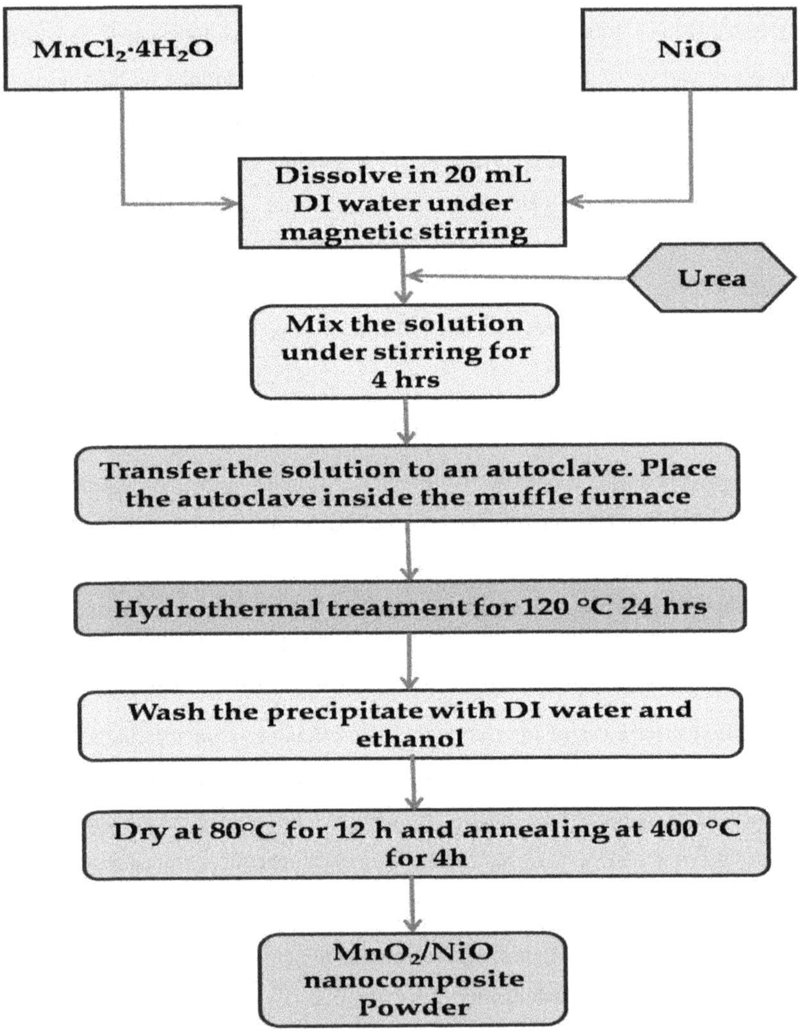

Fig. 2.1: Synthesis flowchart of MnO_2/NiO nanocomposites by simple hydrothermal route

2.3. RESULTS AND DISCUSSIONS

2.3.1 XRD Analysis

The XRD analysis technique is the one most frequently used technique for identifying the atomic and molecular structure of crystalline materials. Fig. 2.2 illustrates the XRD patterns of MnO_2, NiO and MnO_2/NiO nanocomposite. The prominent peaks appear at 2θ position of 28.6°, 37.4°, 42.8°, 56.7°, 59.3°, and 72.4° were indexed to be (110), (101), (111), (211), (220) and (301) crystal planes of tetragonal MnO_2 (JSPDS card no. 81-2261)[Hong-qiang Wang et al. 2017]. The XRD pattern of NiO nanoparticles indicates a crystallized structure with peaks at 2θ = 37.2°, 43.1°, 62.9 °, 75.4°, 79.4° resultant to the diffractions from (111), (200), (220), (311) and (222) and their lattice parameters are well material with those of JCPDS card no. 71-1179 for the standard spectrum of pure NiO.

These results indicate that the products were nano-NiO crystals of cubic structure; they have a high purity and small particle size with a fine crystal phase [Shijin Zhu et al. 2018]. No further peaks were detected, specifying the high degree of phase purity of NiO prepared in this work. The XRD pattern confirmed the cubic structure and lattice constant was calculated as a = b = c = 4.177 Å. The crystallite sizes of the prepared samples were calculated using Scherrer relation. The crystallite size of most intense peak was calculated. To calculate the crystallite size the following relation is used.

$$\mathbf{D = 0.9\lambda/\beta \cos\theta} \qquad \textbf{(2.1)}$$

Where,

D = crystallite size

λ = X-ray wavelength

β = Full width at half maximum

θ = angle

The average crystallite size of NiO and MnO_2 is calculated to be 29.75 nm and 35.89 nm respectively. The formation of nanocomposite of different materials can be identified through XRD analysis. Here, XRD result confirms the successful preparation of MnO_2/NiO composites.

All the diffraction peaks of MnO_2/NiO nanocomposite are established to the NiO with cubic phase and other peaks are corresponding due to MnO_2. The presence of peaks corresponding to both MnO_2 and NiO in the MnO_2/NiO nanocomposites ensures the successful formation of the composite materials. Furthermore, XRD does not detect any signals from other materials, which indicates the high purity of the synthesized product. Besides, the peak located at 37.3° became broad and shifted to a lower degree direction, which may be due to the overlap of (111) plane of NiO and (101) plane of MnO_2. The XRD results point out the coexistence of NiO and MnO_2.

2.3.2 FTIR Analysis

The FTIR spectral analysis is very useful for the structural analysis of metal oxides materials and their composites. The presence of functional groups and the purity of the prepared materials were further confirmed by FTIR spectroscopy. The vibrational atom modes of MnO_2, NiO and MnO_2/NiO nanocomposites were confirmed by FTIR analysis as shown in Fig. 2.3. Peaks observed between 400 and 800 cm^{-1} can be assigned to the metal-oxygen (M-O) stretching vibrations.

From MnO_2 spectra, the wide band around 3432 cm^{-1} is ascribed to O-H stretching vibrations, indicating the hydrates nature of MnO_2. The absorption bands around 1627 and 1036 cm^{-1} are assigned to bending vibrations of O-H combined with the Mn atoms. The peaks at 538 and 434 cm^{-1} are corresponding to Mn-O vibrations. From NiO, a large absorption band at 1637 cm^{-1} is attributable to flexible vibration of H_2O molecules. The band around 1388 cm^{-1} is due to presence of CO_3^{2-} anions. The 424 cm^{-1} strong band corresponds to the vibration of the Ni-O bond. As can be seen, the formation of MnO_2 / NiO nanocomposite is confirmed with two strong bands at 448 and 599 cm^{-1}. These bands indicated the division of the vibratory band corresponding to the Ni-O or O-Mn-O bond [Ma. M et al. 2003]. The peak at 3429 cm^{-1} is due to stretching vibrations of the O-H adsorbed nanocomposite surfaces. The peak at 1629 cm^{-1} are from the vibration of chelators with symmetric and asymmetric COO^- formats.

2.3.3 Morphological Studies

Fig. 2.4 shows the HRSEM images of MnO_2, NiO and MnO_2/NiO nanocomposite and confirmed that all the synthesized samples having the size is non-uniform.

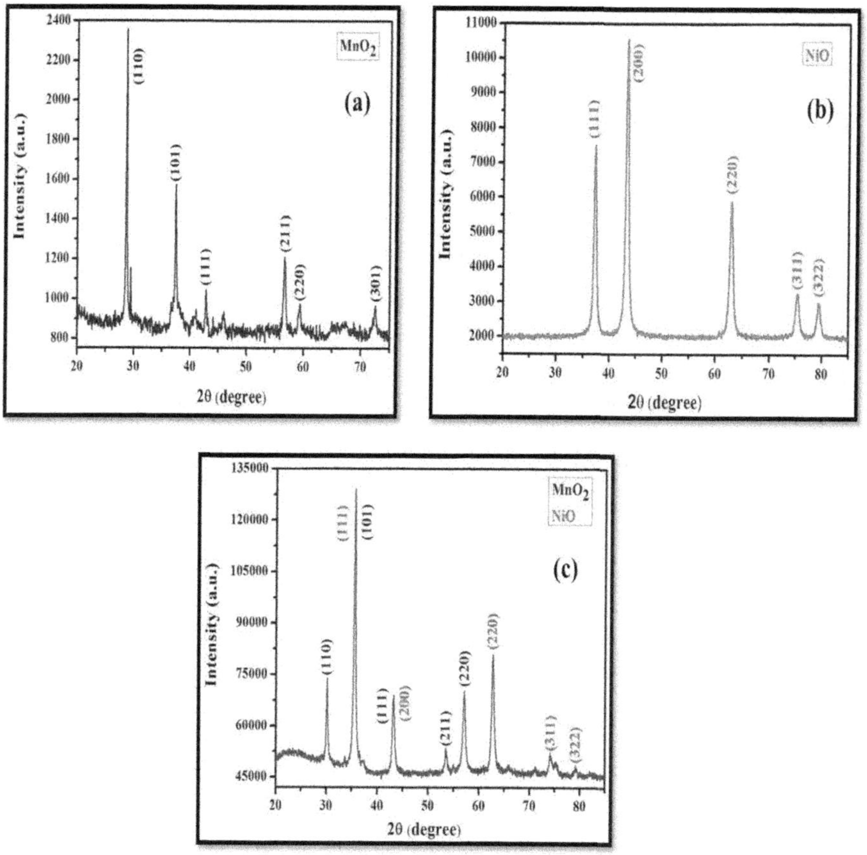

Fig. 2.2: XRD patterns of (a) MnO_2 (b) NiO and (c) MnO_2/NiO nanocomposite.

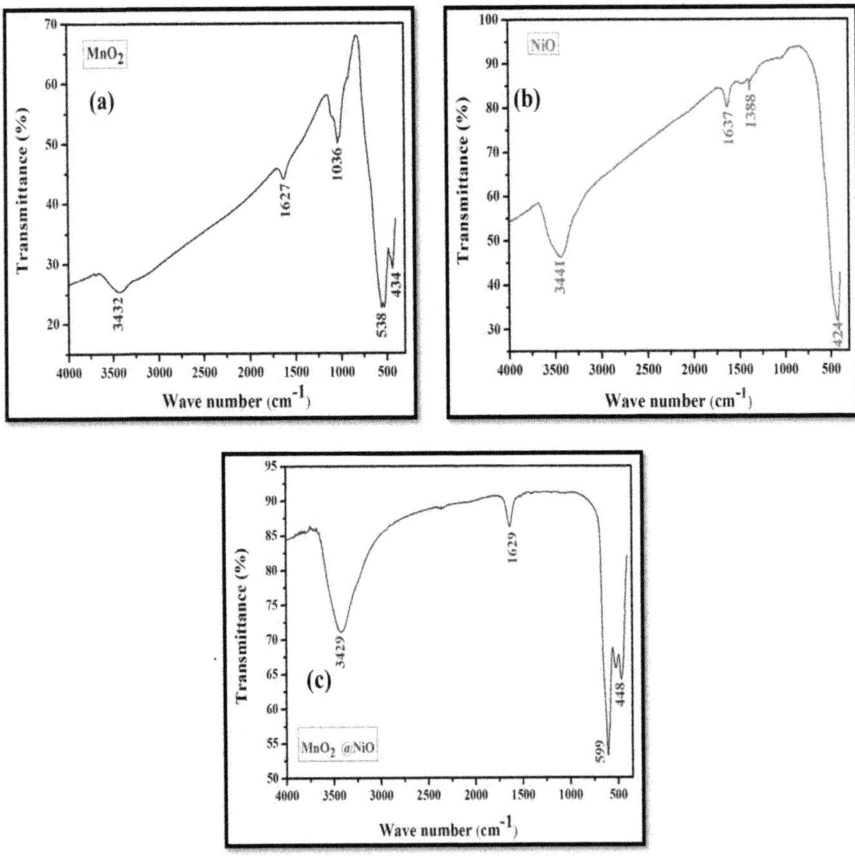

Fig. 2.3: FTIR spectrum of (a) MnO_2, (b) NiO and (c) MnO_2/NiO nanocomposite.

Hence it may be the outcome of the combination of the spherical-shaped particles and the growth of non-equal crystalline grains during the preparation period. Due to low dosage of urea to precursor ratio of 1:1, the number of urea around Ni and Mn was very limited in the solution. In the course of the reaction, as the small amount of urea could not provide sufficient hydroxyl groups, therefore non-uniformly sized spherical like particles was formed, and the shape varied from spherical to rod like morphology after the reaction was carried out for 24h. Similar results were found earlier for CeO_2-ZrO_2 nanocomposites [Xiaohui Zhang et al. 2015]. From the HRSEM images, we concluded that the formation of spherical nanoparticles like structure, due to aggregation followed by the agglomeration of nanoparticles. It was illustrious that the particles are collected together, due to magnetic interaction between the particles [Sonia. M.M.L et al. 2018, Amaliya. A.P et al. 2018].

The elemental composition of electrode material is necessary for the supercapacitor energy storage application; for that reason, elemental and chemical compositions of MnO_2, NiO and MnO_2/NiO nanocomposite were tested by EDX analysis. The EDX spectra of prepared samples were analyzed in binding region of 0-10 keV, which is shown in Fig. 2.5. Fig. 2.5(a) & (b) reveal the presence of stoichiometric proportion of O, Mn and O, Ni in MnO_2 and NiO samples respectively, and Fig. 2.5(c) evidently confirms the presence of Mn, Ni, and O in MnO_2/NiO nanocomposite. The observed atomic weight is confirming the stoichiometric proportion of MnO_2, NiO and MnO_2/NiO nanocomposites (Fig. 2.5 (a-c) inset).

The morphology and structural features of MnO_2, NiO and MnO_2/NiO nanocomposite were also described by HR-TEM analysis. The characteristic HR-TEM images are exposed in Fig. 2.6 (a-c). The results show the average product size of MnO_2 nanoparticles as 10 to 50 nm (Fig. 2.6a) similar to the results from XRD patterns. The uniform MnO_2 particles have spherical shapes with weak agglomeration. According to the TEM image, it can be concluded that this preparation method for the precursor obtained by homogeneous precipitation had successfully overcome the problem of agglomeration and the calcination condition. The observed spherical shaped samples are formed in NiO and MnO_2/NiO nanocomposite by the small aggregation of nanoparticles, due to the magnetic natural surroundings of the materials. The SAED pattern of pristine MnO_2, NiO nanoparticles and MnO_2/NiO nanocomposite were shown in Fig. 2.7(a-c) confirmed the samples were well crystallized in nature.

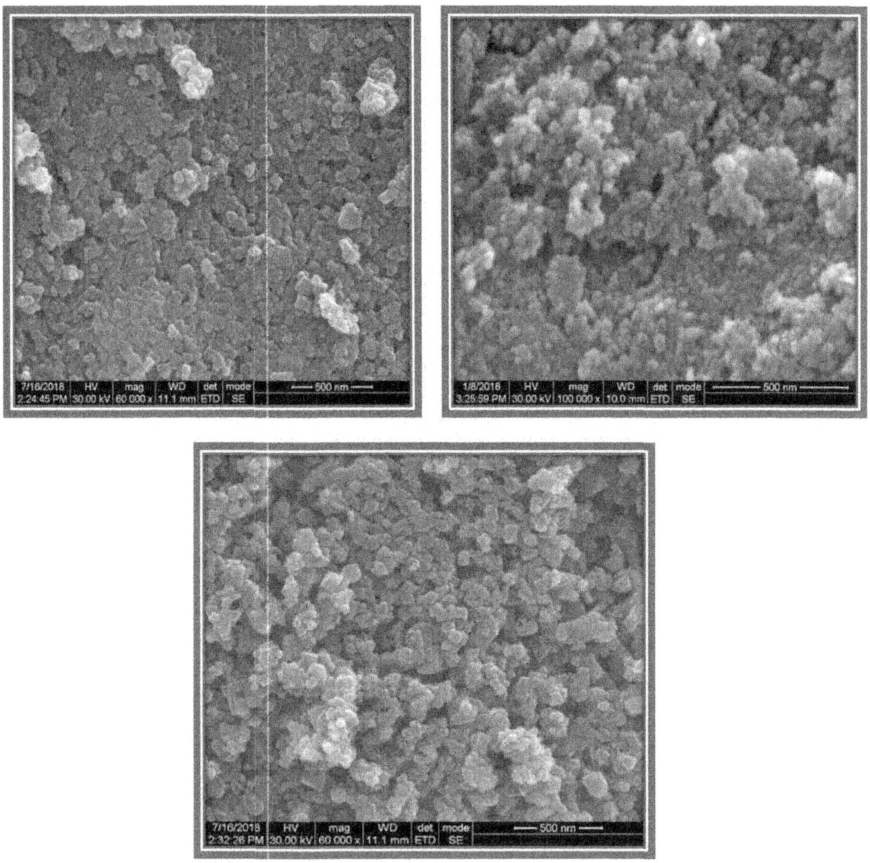

Fig. 2.4: HRSEM images for (a) MnO_2, (b) NiO and (c) MnO_2/NiO nanocomposite.

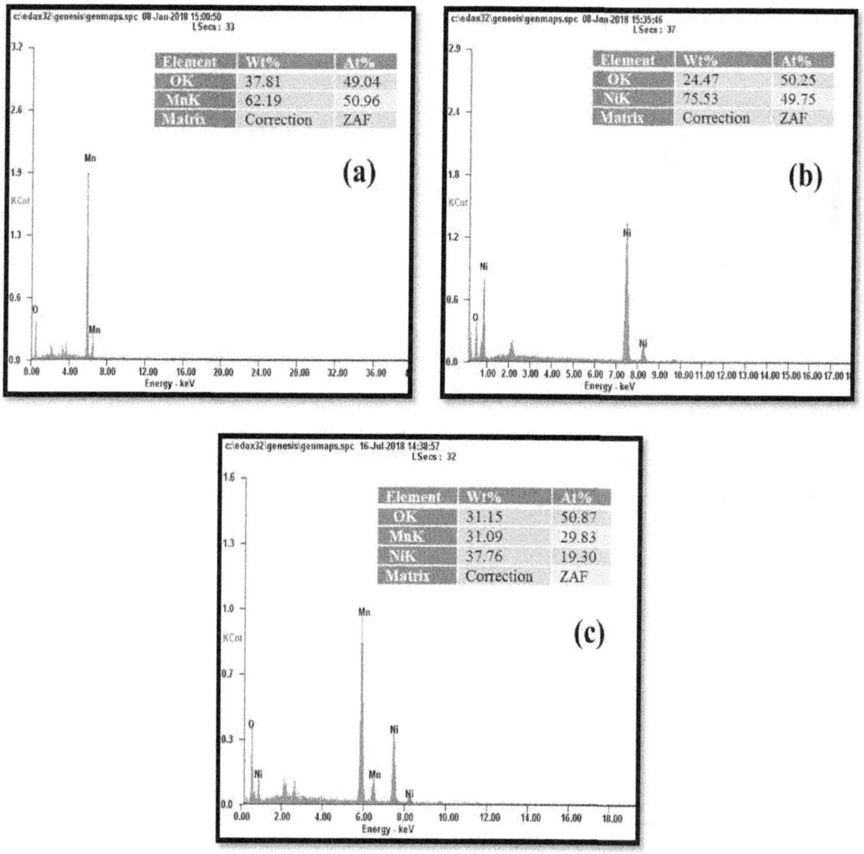

Fig. 2.5: EDX spectrum of (a) MnO_2, (b) NiO and (c) MnO_2/NiO nanocomposite.

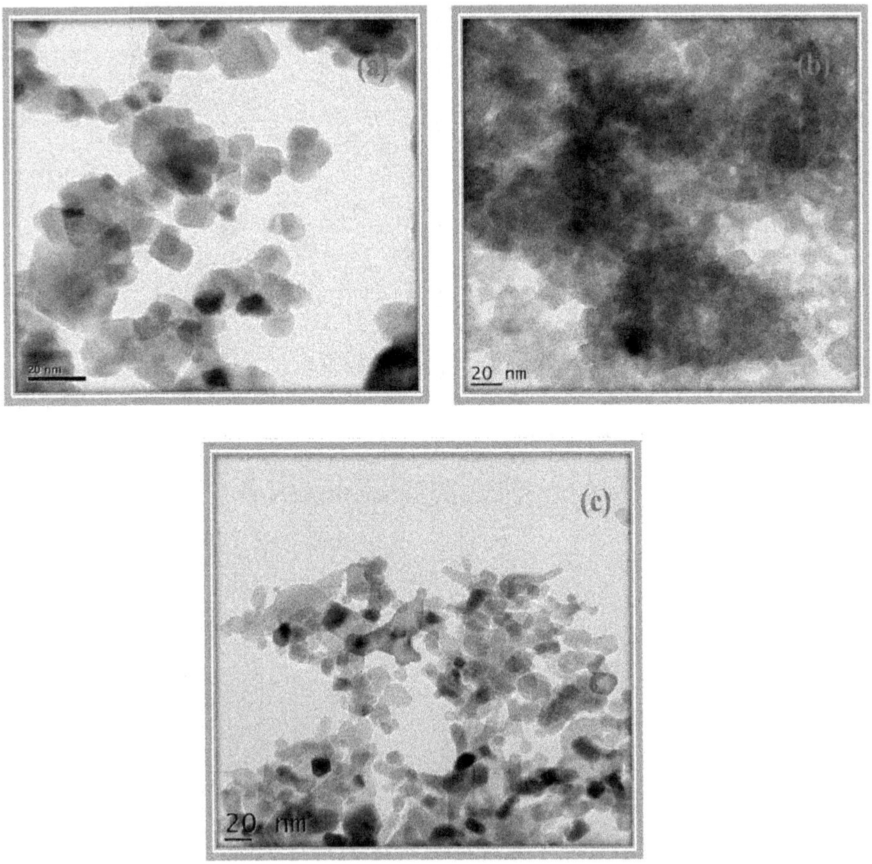

Fig. 2.6: TEM images of (a) MnO_2, (b) NiO and (c) MnO_2/NiO nanocomposite.

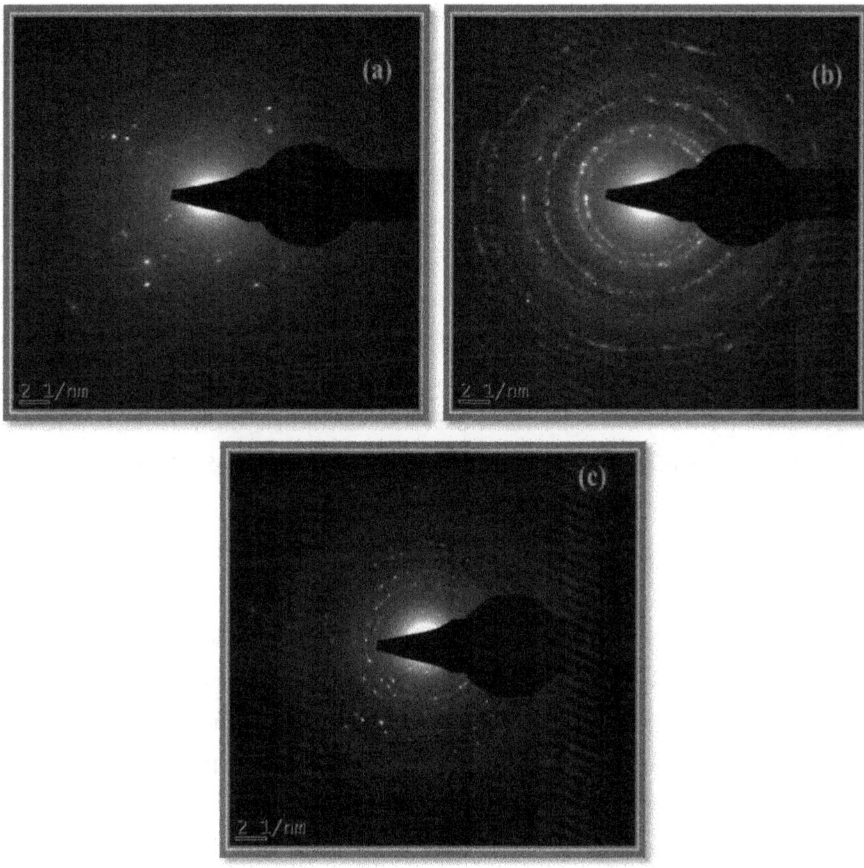

Fig. 2.7: SAED Pattern of (a) MnO_2, (b) NiO and (c) MnO_2/NiO nanocomposite.

2.3.4 VSM Analysis

The magnetic properties of the pristine MnO_2, NiO and composites of MnO_2/NiO composite have been investigated at room temperature in the applied magnetic field of 15000 Oe (Fig. 2.8). Magnetic properties of pristine MnO_2 nanocrystallites depend upon the concentration of magnetic ions, their distribution at different tetrahedral and octahedral sites, and concentration of defect states [Singh. K et al. 2018]. The hysteresis curve of pristine MnO_2 nanocrystallites reveals the ferromagnetic nature in the applied range of magnetic field (Fig. 2.8a). The saturation magnetization, coercively and retentivity values for the pristine MnO_2 nanocrystallites are found to be 0.8 emu/g, 67.91Oe and 0.0074 emu/g, respectively, at the low external applied field. The area enclosed by the hysteresis curve for a ferromagnetic material is an estimation of the energy loss per unit volume over a cycle that is used as information storage capacity for the recording multimedia. The ferromagnetic characteristic of pristine MnO_2 particles can be attributed to the presence of oxygen vacancies, defect state and interaction of Mn^{4+} ions with O^{2-} ions because the magnetic interaction between Mn^{4+}-O -Mn^{4+} is only the ferromagnetic interaction, where as other interaction of Mn ions is antiferromagnetic in nature [Liu. X. M et al. 2005, Masquelier. C et al. 1996]. Thus, the presence of defect states, grain boundaries, their types and concentration may play vital role in the ferromagnetism of MnO_2 nanoparticles.

However, there are some reports suggesting the paramagnetic/ superparamagnetic nature of MnO_2 particles in different phases (α, β, γ and δ) or ferromagnetic/antiferromagnetic behaviour of β-MnO_2 at low temperatures [Kim. H. J et al. 2007, Barudzija. T et al. 2016]. This indicates that MnO_2 can be found in different phases and their magnetic properties depend upon the synthesis process and conditions under which the magnetization measurements are carried out.

The magnetic hysteresis loops of the NiO sample was also measured at room temperature that are shown in Fig. 2.8b. The corresponding values of saturation magnetization (*M*s), coercivity (*H*c) and remanent magnetization (*M*r) are measured as 0.7emu/g, 120.35 Oe and 0.0065 emu/g respectively, It is clear that the ferromagnetic nature was observed as a result of applying external magnetic field. The vibrating sample magnetometry (VSM) curves of MnO_2/NiO composite showing magnetization as a function of applied magnetic field at room temperature are shown in Fig. 2.8c.

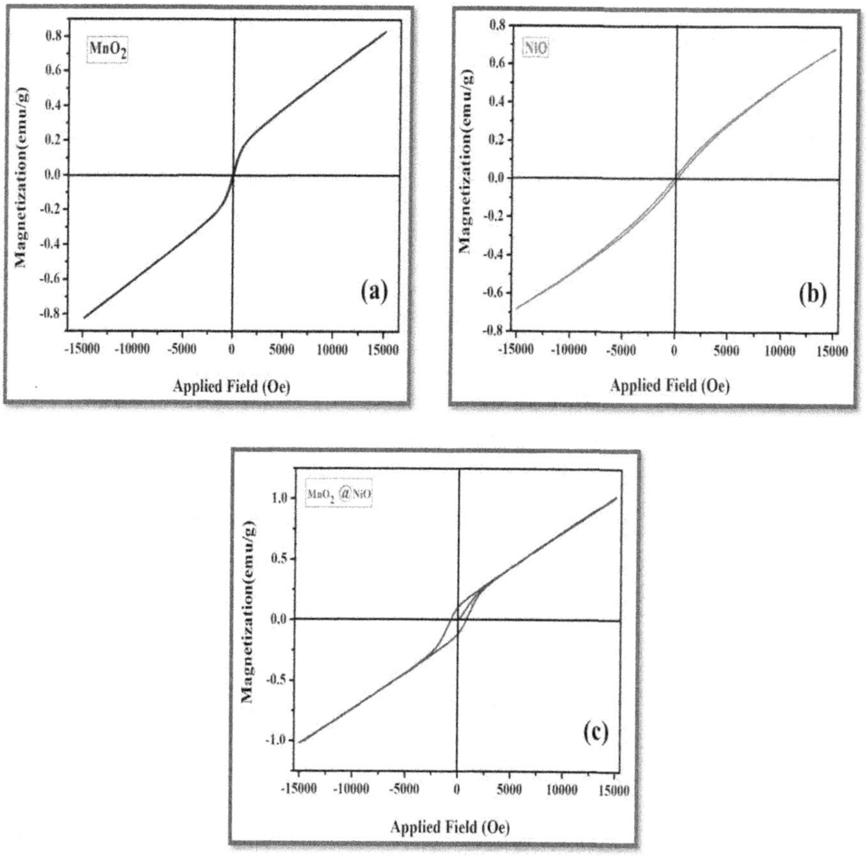

Fig. 2.8: VSM hysteresis loops of (a) MnO_2, (b) NiO and (c) MnO_2/NiO nanocomposites.

A hysteresis loop is observed for the nanocomposites indicating that the nanoparticles could impart ferromagnetic behaviour. It is clear from the loop that the values of magnetic properties such as saturation magnetization (*M*s), remanence (*M*r) and coercivity (Hc) are increased with addition of NiO into MnO_2 metal oxide nanoparticles.

2.3.5 X-ray Photoelectron Spectrum (XPS) Analysis

X-ray photoelectron spectroscopy (XPS) is a quantitative spectroscopic technique that measures the chemical state and elemental composition of the materials. In-depth X-ray photoelectron spectrum (XPS) is utilized to probe the chemical valence states of Ni, Mn, and O in MnO_2/NiO nanocomposite. The survey XPS spectrum (Fig. 2.9a) of MnO_2/NiO confirms the coexistence of Ni, Mn, and O elements. The high-resolution Mn2p XPS region (Fig. 2.9c) consists of two prominent peaks at 652.64 eV and 640.91 eV with 11.7 eV of spin energy separation owing to a spin doublet of Mn $2p_{1/2}$ and Mn $2p_{3/2}$. Moreover, the atom molar rate of Mn and O is about 1:2 in the survey spectra, demonstrating that the realisation of bare MnO_2 and chemical state of manganese element is +4. These results are evidently comparable with the earlier reported data for Mn $2p_{3/2}$ and Mn $2p_{1/2}$ in MnO_2 further declare that the existence of MnO_2 phase in MnO_2/NiO nanocomposite [Liu. Z et al. 2017]. The high-resolution Ni 2p XPS region (Fig. 2.9d) of MnO_2/NiO indicates that nickel elements exist as Ni^{2+}, determined with the Ni $2p_{3/2}$ at 854.34 eV and Ni $2p_{1/2}$ at 872.19 eV [Yang. S et al. 2017, Guo. J et al. 2017]. The corresponding satellite peaks for Ni $2p_{3/2}$ and Ni $2p_{1/2}$ are observed at 861.8 and 879.9 eV, respectively. The higher oxidation state manganese species are more active for electrochemical features over manganese-based electrodes [Zhang. X et al. 2018]. In Fig. 2.9e, the O 1s region reveals a peak at 528.75 eV binding energy due to oxygen species comprising the MnO_2/NiO nanocomposite while a peak existence of oxygen and H_2O molecules.

2.3.6 Electrochemical Measurements

Cyclic Voltammetry (CV) is a widely used technique by electrochemists because of its versatility, but most of the time this technique is devoted to laboratory cells.

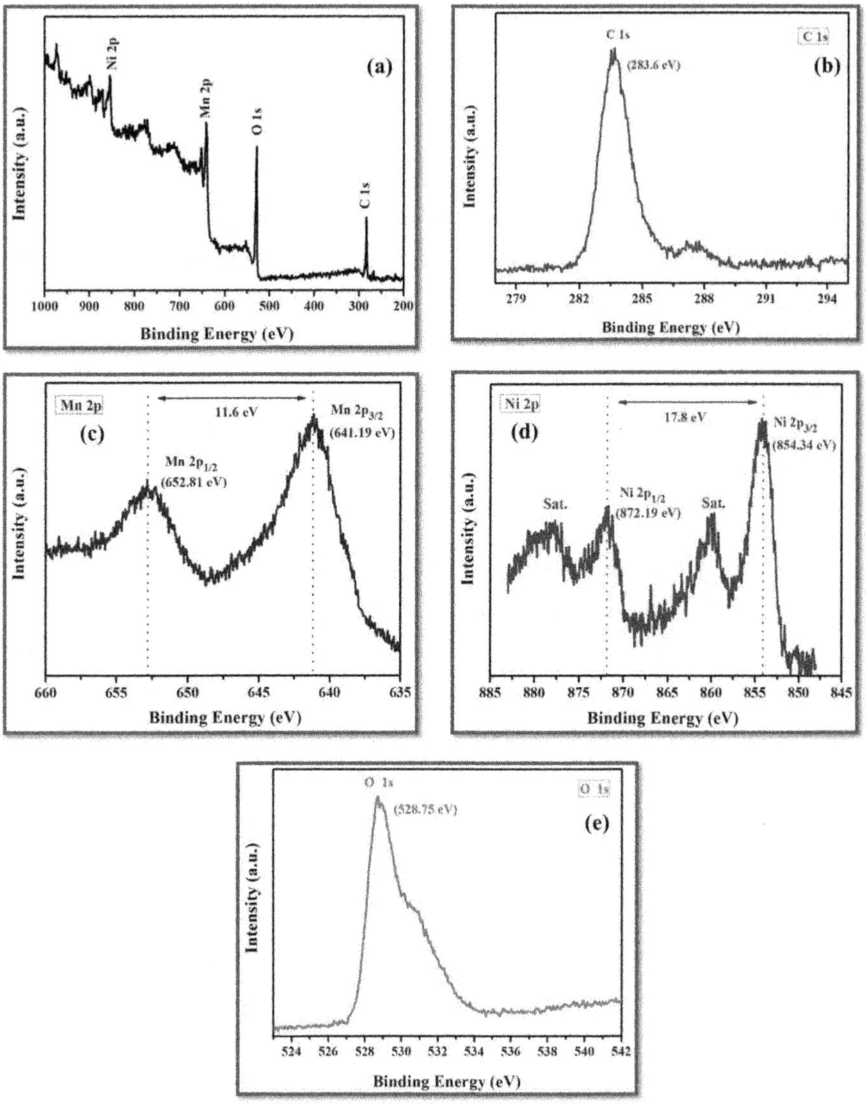

Fig. 2.9: XPS spectra of MnO_2/NiO nanocomposites (a) full XPS survey scan spectrum and (b–e) XPS spectra of C1s, Mn 2p, Ni 2p and O 1s respectively.

At the laboratory or material scale it is an accurate technique that enables:

- qualitative and pseudo-quantitative studies
- kinetic analysis by scanning a huge range of scan rates
- Voltage window determination.

The principle of this technique is to apply a linear voltage ramp to an electrode (or a device) between two voltage limits and to measure the resulting current. The electrochemical performance of the prepared materials was studied through CV, in 1M aqueous KOH electrolyte without any binding or carbon additive. The CV measurements were made in a three-electrode system with the reference electrode (Ag/AgCl) and a counter electrode (Platinum foil). The CV measurement of the electrode was taken in potential range of -0.1 V to 0.4 V. The two regions in CV plot indicate the area of negative current reduction at the cathode and positive current region of anode oxidation.

The CV graphical plots of NiO, MnO_2 and MnO_2/NiO composites are showed in Fig. 2.10. Bare MnO_2 CV curve appears as rectangular shape without obvious redox peaks; capacitive behaviour of MnO_2 nanospheres electrode is pseudocapacitance, attributed to the reversible redox reaction among Mn^{3+} and Mn^{+4}. It was attributed to cation surface adsorption (Ni^{+}) of electrolyte and possible intercalation/ deintercalation of Ni^{+} [Bao. L. H et al. 2011]. With the addition of NiO in MnO_2, the CV hybrid matrix shows a clear difference. The shape of the plot is not perfectly rectangular, which point out a deviation from ideal behaviour of the supercapacitor. The plot contains some redox peaks which are observed in all CV curves.

It suggests the pseudocapacitive characteristic of active materials, which is obviously different from the double layer electrical capacity characterized by almost rectangular CV curves, redox pairs correspond to reversible reactions of Ni^{2+}/Ni^{3+} transitions [Kim. J.H et al. 2010]. The slight change in shape of the CV curves specifies the good charging rate capability of the electrode. Of course, cyclic voltammetry is useful to evaluate the cyclability of a supercapacitor (or an electrode), and the variation in capacitance on cycling would give a good idea of how things will go. But, usually, galvanostatic cycling is better when performing such experiments.

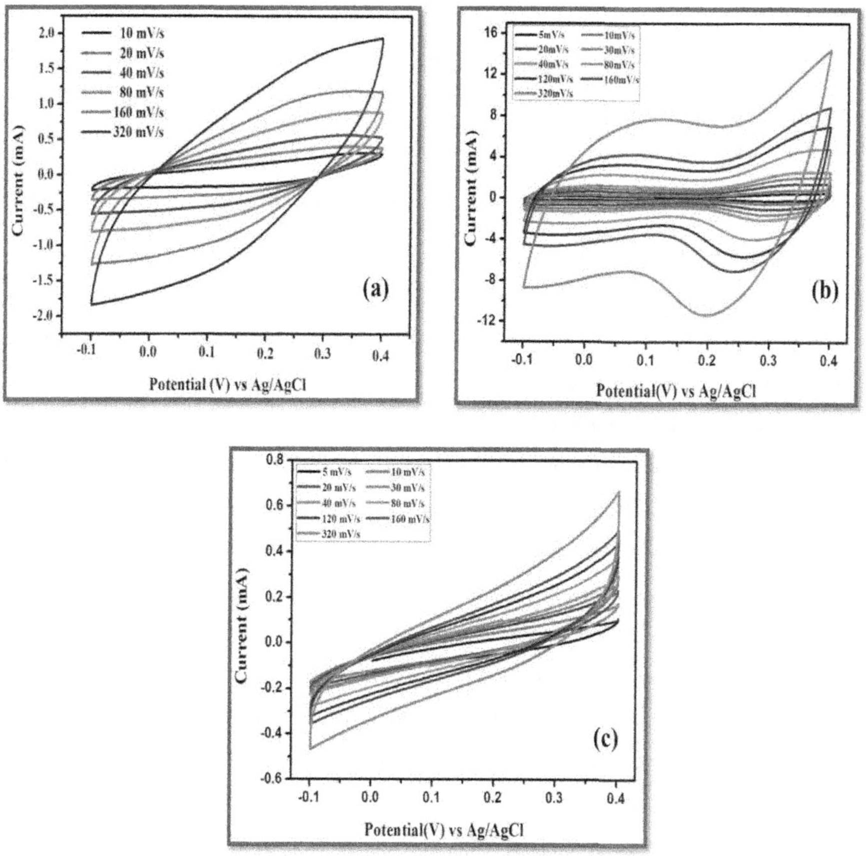

Fig. 2.10: Cyclic Voltammograms at different scan rates of (a) MnO_2, (b) NiO and (c) MnO_2/NiO nanocomposite.

Galvanostatic Cycling technique is very different from cyclic voltammetry because the current is controlled and the voltage is measured. This is certainly the most widely used technique in the supercapacitor field because it can be extended from a laboratory scale to an industrial one. This method is also called chronopotentiometry and gives access to different parameters such as capacitance, resistance, cyclability. The galvanostatic charge–discharge (GCD) test was performed at a current density of 1 Ag^{-1}. The voltage range was maintained from -0.1–0.4 V during the GCD test, as given in Fig. 2.11. The specific capacitance from the GCD plot can be calculated by the equation:

$$\textbf{Specific Capacitance} = C_{GCD} = i \times \Delta t/(\Delta v \times m) \qquad \textbf{(2.2)}$$

Where *i/m* is the current density (Ag^{-1}), Δt corresponds to discharge time and Δv stand for potential range. The nature of the GCD plot deviates from ideal triangular pattern indicating that MnO_2/NiO composite material is pseudocapacitive in nature, which also supports the CV result. As given in Fig. 2.12, contrasting with specific capacitances of MnO_2 and NiO and MnO_2/NiO electrodes at various charge/discharge current, the capacitance of MnO_2/NiO reduces from 247.78 Fg^{-1} (current density of 0.5 Ag^{-1}) to 56.21 Fg^{-1} (4 Ag^{-1}) with 22% of the specific capacitance remaining, in the same process, the capacitance of MnO_2 electrode reduces from 175.51 Fg^{-1} (0.5 Ag^{-1}) to 37.47 Fg^{-1} (4 Ag^{-1}) with 21% of the specific capacitance remaining, and the capacitance of NiO electrode reduces from 129.65 Fg^{-1} (0.5 Ag^{-1}) to 23.72 Fg^{-1} (4 Ag^{-1}) with 17% of the specific capacitance remaining. This result points out the specific capacitance of MnO_2/NiO electrode is much greater than pristine NiO and MnO_2 electrode nanoparticles. The higher specific capacitance of MnO_2/NiO electrode is due to NiO spherical nanoparticles were well wrapped on the surface of MnO_2 nanoparticles to provide composites with larger surface area. Our material also showed a better specific capacitance than the specific capacitance of MnO_2 electrode based composites.

The assessment of energy and power density provided by EC supercapacitors is desirable for any application, particularly for those requiring high loads in short time intervals. Specific power density (W kg^{-1}) plotted versus the specific energy density (Wh kg^{-1}), termed the ''Ragone plot'', is often used to evaluate and compare the respective performance of electrochemical energy storage devices.

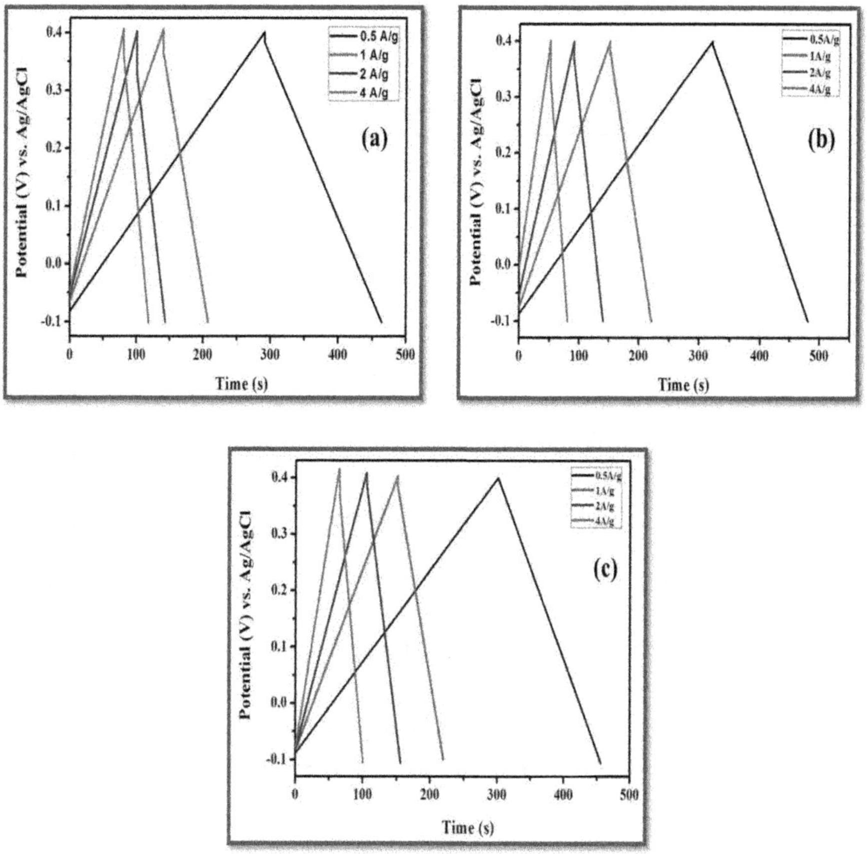

Fig. 2.11: Charge–discharge curves of (a) MnO_2, (b) NiO and (c) MnO_2/NiO nanocomposite at different current densities.

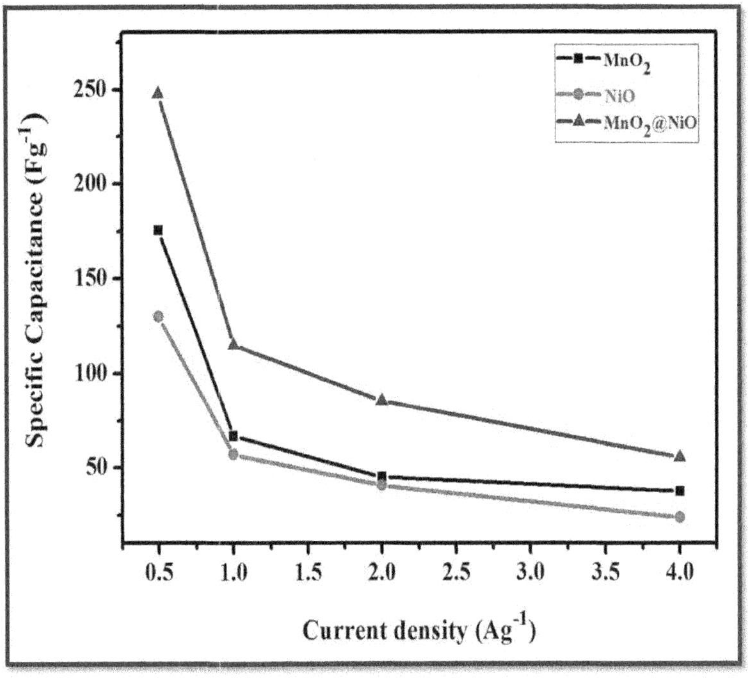

Fig. 2.12: Specific Capacitance (Cs) versus discharge current densities of MnO_2, NiO and MnO_2/NiO nanocomposite electrodes.

The relation between the energy density (E with respect mass or volume) and power density is important to evaluate in order to understand the relative performance between these two properties for practical application. The power density defines the rate of power delivered by a device of certain mass at a specified current density. Power densities are influenced by kinetic polarization at over potentials, Ohmic resistances of the electrolyte, and intrinsic material structure. The energy density (E_d) values of the prepared electrode materials and power density (P_d) values were calibrated by following relations.

$$\mathbf{E_d = 0.5C_s\Delta V^2} \tag{2.3}$$

$$\mathbf{P_d = \frac{E_d}{t}} \tag{2.4}$$

Where, E_d is energy density (Wh kg^{-1}), C_s is specific capacitance (Fg^{-1}), ΔV is potential window of discharge (V), P_d is power density (W kg^{-1}) and Δt is discharge time (s). The energy density is found to increase with the power density. The power density is in increasing trend which point out the pseudocapacitor discharging capacity is directly proportional to discharge current. At a current density of 0.5 Ag^{-1}, the MnO_2/NiO nanocomposites provided a higher energy density of 971.2948 Wh kg^{-1} and power density of 0.1346 kW kg^{-1}. Table 2.1 exhibits the specific capacitance, energy and power density values of MnO_2, NiO and MnO_2/NiO nanocomposites. Table 2.2 presents the supercapacitor performance comparisons of MnO_2/NiO nanocomposites prepared in the present work revealed better cycling performance than some of the earlier reports. From the GCD plot, the Coulombic efficiency was calculated by the following equation:

Coulombic Efficiency= [discharging time (t_D)/charging time (t_C)]×100%. (2.5)

The Coulombic efficiency (calculated with respect to the second GCD cycle) of MnO_2/NiO is 104%, which is good for potential supercapacitor applications. The GCD test was continued up to 10000 cycles.

Table 2.1: Specific Capacitance, Energy and Power density values of MnO_2, NiO and MnO_2/NiO Nanocomposite

Sample	Current density (Ag^{-1})	Specific Capacitance (Fg^{-1})	Energy density ($Wh \cdot kg^{-1}$)	Power density ($kW \cdot kg^{-1}$)
MnO_2	0.5	175.51	688.0270	0.1508
	1	66.78	261.7838	0.3016
	2	45.15	176.6254	0.6031
	4	37.47	146.8871	1.2062
NiO	0.5 A	129.65	508.2483	0.2734
	1	57.10	223.4850	0.5468
	2	40.55	158.9626	1.0937
	4	23.72	92.9985	2.1874
MnO_2/NiO	0.5	247.78	971.2948	0.1346
	1	114.76	449.8592	0.2562
	2	85.24	334.1596	0.5124
	4	56.21	217.1993	1.0248

Table 2.2: Comparison of Maximum Capacitance values of the reported and the Present Electrode of MnO_2/NiO Nanocomposite

Electrode Material	**Morphology**	**Electrolyte**	**Capacitance ($F\ g^{-1}$)**	**Reference**
MnO_2-Based Mixed Oxides	Spherical	Na_2SO_4 (1M)	210 Fg^{-1} (0.12Ag^{-1})	[Kim. H and Popov. B.N 2003]
$NiMn_2O$	Nanosheets	Na_2SO_4 (1M)	370.5 Fg^{-1} (1Ag^{-1})	[Hongmei Wei et al. 2016]
NiO@MnO_2	Nanosheets	KOH(2M)	266.7 Fg^{-1} (0.5Ag^{-1})	[Chen. J et al. 2016]
α-MnO_2 /graphene	Nanorod	Na_2SO_4(0.5M)	360Fg^{-1} (2.5Ag^{-1})	[Haldorai. Y et al. 2016]
MnO_2/graphene	Honeycomb	KOH(1M)	278.5 Fg^{-1} (1Ag^{-1})	[Zhong. J et al. 2017]
NiCo-MnO_2 nanocomposite	Spherical	KOH(1M)	264.1 (0.5 Ag^{-1})	[Oyedotun. K.O et al. 2018]
Manganese-Nickel Oxides	Spherical	Na_2SO_4 (0.1M)	160 Fg^{-1}	[Chen. Y and Hu. C 2003]
Nano sized Ni-Mn Oxides	Spherical	KOH (4 M)	198.8 Fg^{-1}	[Zhou. J et al. 2006]
MnO_2@NiO nanocomposites	**Spherical**	**KOH(1M)**	**247.7 Fg^{-1} (0.5 Ag^{-1})**	**This Work**

To know the cyclic stability of MnO_2/NiO, we have plotted the specific capacitance vs. the number of cycles and it is shown in Fig. 2.13. After 10000 cycles, the specific capacitance remains stable and the retention of specific capacitance is 81.2%, which is also excellent for supercapacitor applications.

Electrochemical Impedance Spectroscopy (EIS) allows covering a large time scale (from microseconds to hours), meaning that electrochemical processes are parted depending on their own time constant. In addition, such measurements are made under steady-state conditions that allow acquisition times high enough to get accurate measurements. At last, in contrast to the previous techniques, small excitation signals are employed, permitting linearization of current–voltage characteristics. The kinetics of the above prepared pristine and composite electrodes materials were examined by EIS analysis. The Fig. 2.14 signifies Nyquist plots of the MnO_2, NiO and MnO_2/NiO nanocomposite. EIS of pristine MnO_2, NiO and MnO_2/NiO nanocomposite were taken in open circuit potential at a frequency sequence of 0.1 Hz to 100 kHz with 10 mV of functional potential amplitude. The Nyquist plot indicates in the high-frequency interval a single semicircle and the low-frequency interval in a linear peak. The arc indicates double-layered capacitance and charge transmission resistance. The occurrence of the point is acknowledged to dispersal of ions neighboring the electrode surface, demonstrating the pseudocapacitive properties of the MnO_2, NiO and MnO_2/NiO nanocomposite. The semicircle length denotes the charge transfer resistance, (R_{ct}=~2Ω) in the electrochemical technique. The EIS results validate the CV results specifying that the extreme capacitance participation of the composite electrode material is established on the fast surface redox reaction. All of the results indicate that the MnO_2/NiO nanocomposite is a suitable electrode material for supercapacitors.

2.4 CONCLUSION

MnO_2 based composite materials have many advantages, which makes them promising materials for advanced SCs. It seems that SC electrode materials based on MnO_2 will occupy more and more prominent position in the research of SCs. In conclusion, the MnO_2/NiO nanocomposite was successfully prepared by hydrothermal method with cost effective precursors to investigate the electrochemical performance of supercapacitors.

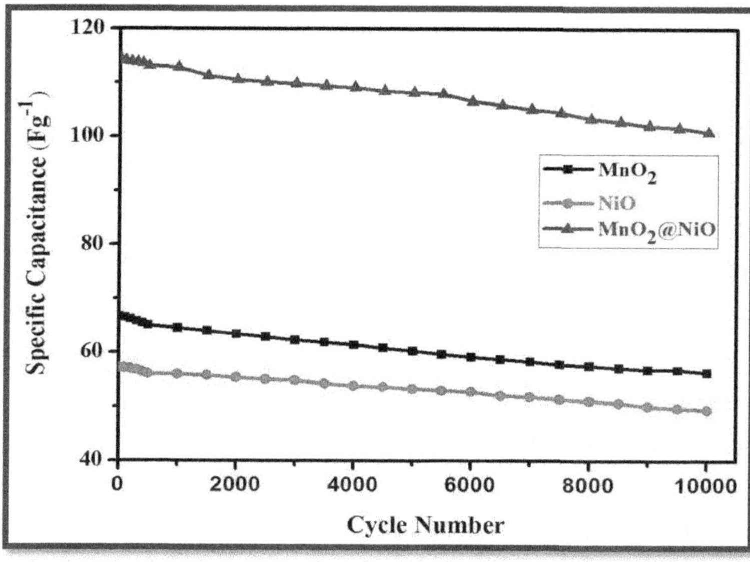

Fig. 2.13: Cycling performance of the MnO_2, NiO and MnO_2/NiO nanocomposite electrodes.

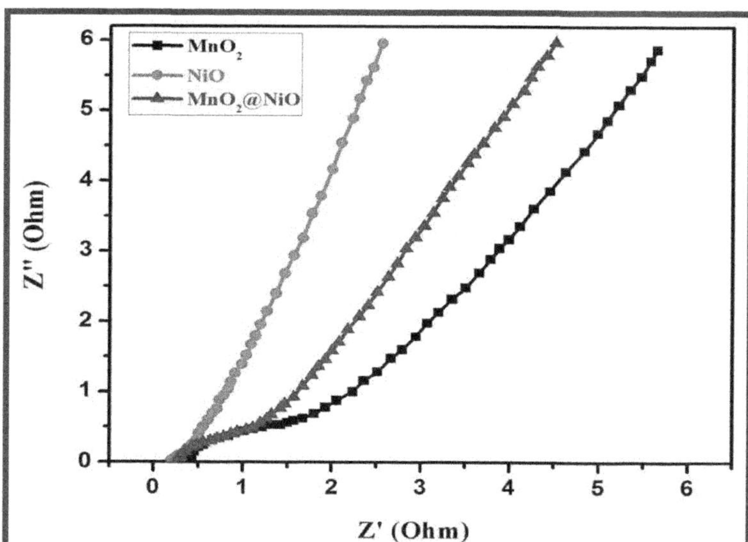

Fig. 2.14: Nyquist plots of MnO_2, NiO and MnO_2/NiO nanocomposite electrodes.

The easy availability, low cost of the reactants and the mild hydrothermal conditions make this method an economic and simple way to produce these materials. The synthesized MnO_2/NiO nanocomposite was characterized using XRD, HRSEM, EDX, TEM, XPS and VSM analysis. The synthesis of the MnO_2/NiO nanocomposite has been confirmed with FT-IR spectroscopy and X-Ray diffraction studies. XRD study on the MnO_2/NiO nanocomposite particles reveals the polycrystalline nature which was confirmed in SAED Pattern. The surface morphological analysis of all the prepared samples revealed the spherical shaped nanocrystalline of the samples. The XPS technique was applied to determine the elemental composition of MnO_2/NiO nanocomposites. The result indicates the presence of Ni, Mn, and O in the nanocomposite sample and no additional elements were observed. The VSM studies performed on the MnO_2/NiO nanocomposite suggest the ferromagnetic nature of the particles at room temperature. The ferromagnetic behaviour of the pristine MnO_2 nanocrystallites can be ascribed to the magnetic interaction between Mn^{4+}-O_2 -Mn^{4+}. The electrochemical achievements of prepared electrodes have been studied through cyclic voltammetry (CV) and galvanostatic charge/discharge test. The MnO_2/NiO nanocomposite reveals a stable cycle performance and notable rate performance. The structural and surface morphology studies of MnO_2/NiO nanocomposite have revealed nanocrystalline nature. This increased conductivity and uniform morphology of the compound accelerates transfer of electrons and the electrolyte diffusion in the hybrid compounds. The MnO_2/ NiO nanocrystalline composite shows a high specific capacity of 247 Fg^{-1} at a current density of 0.5 Ag^{-1}, energy density of 971.2948 $Wh \cdot kg^{-1}$ and power density of 0.1346 kW kg^{-1} compound with good rate capacity and cyclic stability. Improved electrochemical performance of MnO_2/NiO than pristine MnO_2 and NiO could be considered as promising potential electrode materials for the fabrication of stable, cost-effective, and high-performance next generation supercapacitors.

CHAPTER III

SYNTHESIS AND CHARACTERIZATION OF MnO_2/CuO NANOCOMPOSITES FOR SUPERCAPACITOR APPLICATION

3. 1. INTRODUCTION

Electrochemical capacitors (ECs) or Supercapacitors (SCs) is a type of energy-storage devices, and have become an important research topic due to their attractive features such as fast charge/discharge rate, high energy, power density, and very long cycle life [Cheng. Y et al. 2013, Bao. L et al. 2011, Ju. H.F et al. 2014]. The high performance of supercapacitors is closely correlated to their energy storage principle, which could be divided into two types: electrochemical double layer capacitance (EDLC) and pseudocapacitance [Winter. M and Brodd. R.J 2004]. The EDLC's storage mechanism is connected with electrostatic interaction between ions on the surface of electrode material and the electrolyte, i.e. the rapid adsorption/desorption process [Jost. K et al. 2014]. The pseudocapacitance mechanism is achieved by the rapid reversible redox reaction of the electrode surface and doping/undoping process of ions to complete the energy storage and release [Winter. M and Brodd. R. J 2004, Jost. K et al. 2014].

The energy storage principle of the SCs mainly depends on the choice of the electrode material. The parameters such as the specific capacity, energy density and power density of the supercapacitor are determined by the performance of electrode material [Cao. M et al. 2007]. Currently used supercapacitor electrode materials are mainly carbon materials [Hou. J.H et al. 2015], metals [Wang. Q.F et al. 2017], transition metal oxides [Wang. L et al. 2013], conducting polymers [Zhao. D.P et al. 2019] etc. Among them, transition metal oxides can provide a variety of oxidation states for efficient redox charge transfer, leading to much higher specific capacitance [Li. Q et al. 2012]. MnO_2 materials are promising pseudocapacitance materials because of their high specific capacitance, ideal capacitor charge–discharge curves, abundantly of materials, and environmental friendliness [Sun. Z et al. 2013].

MnO_2 is a stable manganese oxide, it has a theoretical high specific capacitance of up to 1370 Fg^{-1}, but in practical use, the capacitance is far less than this theoretical capacitance value [Liu. L et al. 2016, Toupin. M et al. 2004] have shown that the performance of manganese oxide pseudocapacitor is mainly limited by the thickness of the surface layer. The surface layer has a minimum thickness of 420 nm, which causes the migration of electrolyte ions into the manganese oxide layer to a certain extent. Only part of the manganese oxide is fully utilized. MnO_2 is a semiconductor material: it not only has double capacitance properties, but also pseudocapacitance properties. The energy storage of this material is based on the charge in the surface desorption, the occurrence of redox reaction processes. In 1999, Lee and Good enough made a pioneering study of the pseudocapacitive behavior of manganese oxides in aqueous solutions. The pseudocapacitive reaction that occurs on the surface and the main body of the electrode is the main charge storage mechanism of the manganese oxide. It mainly contains two kinds of charge and discharge mechanisms: one is a metal cation in the electrolyte, through the surface of the manganese oxide adsorption/desorption process; the other is the electrolyte in the manganese oxide near the surface of the material and the somatic region of the embedding/de-embedding process. Manganese oxide is among the interesting class of materials, which exist in various forms such as MnO, Mn_3O_4, Mn_2O_3, and MnO_2 as stable oxides. According to literature studies, the crystallinity, crystal structure, morphology, and specific surface area are key parameters which decide the performance of Mn-based oxides. It is well known that high crystallinity of material favors enhanced conductivity. Huang et al. (2014) fabricated porous MnO_2 nanotubes of 6 nm thickness using template-assisted one-step hydrothermal approach. It exhibited specific capacitance of 365 Fg^{-1} at 0.25 Ag^{-1} in a three-electrode system. The asymmetric device was fabricated using MnO_2 nanotubes as positive electrode and activated graphene as negative electrode and it delivered specific capacitance of 50 Fg^{-1} at a current density of 0.25 Ag^{-1}, and cyclic stability with capacity retention of 76.3% after 10000 cycles. The device also delivered an energy density of 22.5 Wh kg^{-1} with maximum power density of 146.2 kW kg^{-1}. But some drawbacks of MnO_2 such as poor electron conductivity and instability restrict the practical application for these supercapacitors [Subramanian. V et al. 2006].

In order to overcome these shortcomings, enormous efforts have been made in the past decades. To overcome this restriction and enhance the electrochemical performances, one effectual strategy is to introduce another transition-metal compound nanostructure on MnO_2 to assemble the nanocomposites. On the other hand, the Copper oxide may also be considered as potential material for supercapacitor application which is also active materials to contribute pseudocapacitive can provide a high specific surface area and highly conductive, ecological friendliness, low cost [Mishra. A.K et al. 2018, Su. X et al. 2014]. Wang et al. (2018) developed electrode composed of CuO nanoparticles by employing oxidation assisted dealloying technique and subsequent electrochemical reactions. The electrode material displayed 210.9 Fg^{-1} and 163.3 Fg^{-1} at 0.5 Ag^{-1} and 10 Ag^{-1} current densities in three-electrode system and offered capacity retention of 94.5% after 5000 cycles at 3 Ag^{-1} current density. The assembled asymmetric supercapacitor device showed long-term cycling stability with more than 93.3% retention after 5000 cycles. It also exhibited high energy density of 20.04 Wh kg^{-1} and at a power density of 7.1 kW kg^{-1}.

Furthermore, the synergetic effect is also playing an important role to improve the performance for the MnO_2-CuO nanocomposite. Therefore, the MnO_2 material and high conductivity CuO material composite can improve the conductivity, electrochemical performance, cycling stability, and other important properties than would be the case for each individual oxide material. Herein, we present a MnO_2-CuO nanocomposite as a pseudocapacitive electrode prepared material by using hydrothermal process. The Crystalline structure, morphology and surface state properties of MnO_2-CuO nanocomposite have been investigated in detail. The electrochemical properties of the MnO_2-CuO nanocomposite were investigated using CV, GCD, and EIS techniques. The construction of the nanocomposite enhances the specific capacitance and long-term cycling stability which is suitable for the SCs application.

3.2. EXPERIMENTAL

3.2.1 Preparation of MnO_2 Nanoparticles

Firstly, MnO_2 nanoparticles were prepared using hydrothermal process using the precursor of $MnCl_2 \cdot 4H_2O$ (Manganese Chloride). The metal precursor and CH_4N_2O were mixed in DI water to obtain a clear solution. Then obtained solution was completely transferred into closed autoclave setup and treated as hydrothermal approach at 120°C for 24 hours. The precipitates obtained from above process were washed with an assistance of DI water and methanol. Finally, the resulting sample was dried at 80° C for 12h and thereafter calcinated at 400°C for 3 h to get MnO_2 nanoparticles.

3.2.2 Preparation of CuO Nanoparticles

The CuO NPs was also obtained using same hydrothermal route without any additional surfactants. First, 4 g copper chloride precursor was mixed in 50 mL DI water, and then 40 mL of NaOH aqueous solution was added into above aqueous solution under continuous stirring. Then, above solution was shipped into sealed autoclave and treated at 110°C for 24 h. The precipitates achieved from above treatment was washed using DI water and methanol. Thereafter, the sample was dried out at 90°C and annealed at 400°C for 3 h in furnace to obtain the crystalline CuO NPs.

3.2.3 Preparation of MnO_2/CuO Nanocomposites

For the MnO_2/CuO nanocomposite preparation, the above synthesized MnO_2 was dissolved in 50 mL double distilled water under magnetic stirring, then $CuCl_2 \cdot 2H_2O$ and NaOH was mixed into the dispersion. After constantly stirring for 1h, then the resulted solution was shifted into a 150 mL autoclave treated at 120°C for 24 h. The settled precipitate for MnO_2/CuO nanocomposite powders was centrifuged using with DI water and methanol. The washed product was dried at 90°C for 24 h and sintered at 400°C for 4h.

3.2.4 Characterizations

The structural nature of synthesized pristine particle and nanocomposites were confirmed by powder XRD analysis using Rigaku D/max-kA diffractometer with CuKα (λ=1.54056 nm).The FT-IR study was performed in frequency range from 4000-400 cm^{-1} (Perkin-Elmer spectrometer) to confirm functional groups and co-ordination. The morphological images of resultant composite and their individual compounds were recorded by scanning electron microscopy (HRSEM, FEI Quanta FEG 200) which is equipped with EDX for elemental composition and the TEM images were obtained on a Philips CM200FEG field emission microscope. The surface electron state and composition of the MnO_2@CuO composites were evaluated by XPS instrument namely PHI 5000 Versa Probe III Photo electron spectrometer.

3.2.5 Electrochemical Measurements

Electrochemical properties of bare MnO_2, CuO and MnO_2/CuO nanocomposite were inspected by cyclic voltammetry (CV), Galvanostatic Charge and Discharge (GCD) and Electrochemical Impedance Spectroscopy (EIS). Working electrode was made of active material (80% by weight), activated carbon (10% by weight) and polyvinylidene fluoride suspension (PVDF) (10% by weight) ratio using N-methyl-2-pyrrolidone (NMP) as a solvent to make homogeneous slurry. After the slurry was pasted onto the nickel sheet current collector and dried at 80°C for 24h. All measurements were taken using an aqueous 1M KOH as electrolyte at ambient conditions with the electrode material prepared as a working electrode, a platinum wire as a counter electrode and Ag/AgCl as a reference electrode.

3.3. RESULTS AND DISCUSSIONS

3.3.1 Powder XRD Analysis

The structural analysis of the prepared samples has also been done using X-ray diffraction studies. Fig. 3.1 shows the X-ray diffraction (XRD) patterns of MnO_2, CuO and MnO_2/CuO nanocomposite. Fig. 3.1(a) shows the XRD patterns of the MnO_2 NPs. The diffraction peaks such as (110), (101), (111), (211), (220) and (301) were observed at 28.6°, 37.4°, 42.8°, 56.7°, 59.3°, and 72.4°, respectively. The observed XRD pattern is well matched with β-phase of MnO_2 (JCPDS No. 81-2261) [Ammundsen. B and Paulsen. J 2001] that confirm the formation of MnO_2 particles in β-phase and tetragonal structure of the crystals with *P42/mnm* space group. Moreover, extra peaks are not observed related to the impurity phases such as α, γ, and δ-MnO_2. The crystallite size of the MnO_2 particles have been calculated from the sharp reflection (110) using well known Scherrer's formula,

$$\mathbf{D = 0.9\lambda/\beta \cos \theta} \qquad \textbf{(3.1)}$$

Where,

D = crystallite size

λ = wavelength of X-ray

β = full width at half maxima

θ = angle

The crystallite size evaluated using Scherrer's formula found to be 35.89 nm, which suggests the nanocrystalline nature of the MnO_2 particles. Fig. 3.1(b) shows the XRD profiles of pure copper oxide (CuO). The diffraction peaks such as (110), $(\bar{1}11)$, (111), $(\bar{2}02)$, (020), (202),$(\bar{1}13)$, $(\bar{3}11)$, (220), (311) and $(\bar{2}22)$ were found at 2θ of 32.5°, 35.6°, 38.8°, 48.9°, 53.6°, 58.4°, 61.7°, 66.2°, 68.2°, 72.5° and 75.2°. All these diffraction peaks were exactly coincided with that of monoclinic structure of CuO (JCPDS card no. 89-5895) [Devi. L.V et al. 2018]. No additional impurity peaks were observed, declares the high degree of phase purity of CuO prepared in this work. The average crystallite size was calculated using the above mentioned Scherrer's equation found to be CuO is calculated to be 26.8 nm. Fig. 3.1(c) presents the XRD patterns of MnO_2/CuO nanocomposite. All the diffraction peaks of MnO_2/CuO nanocomposite are established to the MnO_2 with cubic phase and other peaks are corresponding due to CuO. The XRD results point out the coexistence of MnO_2 and CuO.

3.3.2 FT-IR Analysis

The molecular structure and functional groups of the prepared MnO_2, CuO NPs and MnO_2/CuO composite samples were investigated by FTIR studies. Fig. 3.2 displays the FTIR spectra of the bare MnO_2, CuO NPs and MnO_2/CuO composite synthesized by hydrothermal method which was acquired in the range of 400–4000 cm^{-1}. The peaks observed around 536 and 431 cm^{-1} are ascribed to Mn-O vibrations. The wide band around 3438 cm^{-1} is attributed to O-H stretching vibrations, indicating the hydrates nature of MnO_2. The absorption bands around 1643 and 1040 cm^{-1} are indexed to bending vibrations of O-H combined with the Mn atoms. The peak noted at lower wavenumbers (443 cm^{-1}) confirmed the Cu-O stretching [Sankar. R et al. 2014]. The occurrence of OH groups in the CuO NPs was confirmed by the peaks observed at 3451 and 3321 cm^{-1} and it was attributed to presence of water. The formation of MnO_2/CuO nanocomposite is confirmed with two strong bands at 522 and 451 cm^{-1}. These bands indicated the division of the vibratory band corresponding to the Cu-O or O-Mn-O bond [Padil. V.V.T and Černík. M 2013]. Peak related to Cu-O stretching mode of vibration can be observed in the FTIR spectrum of MnO_2/CuO nanocomposite sample along with Mn-O vibration peak which results the successful formation of nanocomposite and agree well with XRD results.

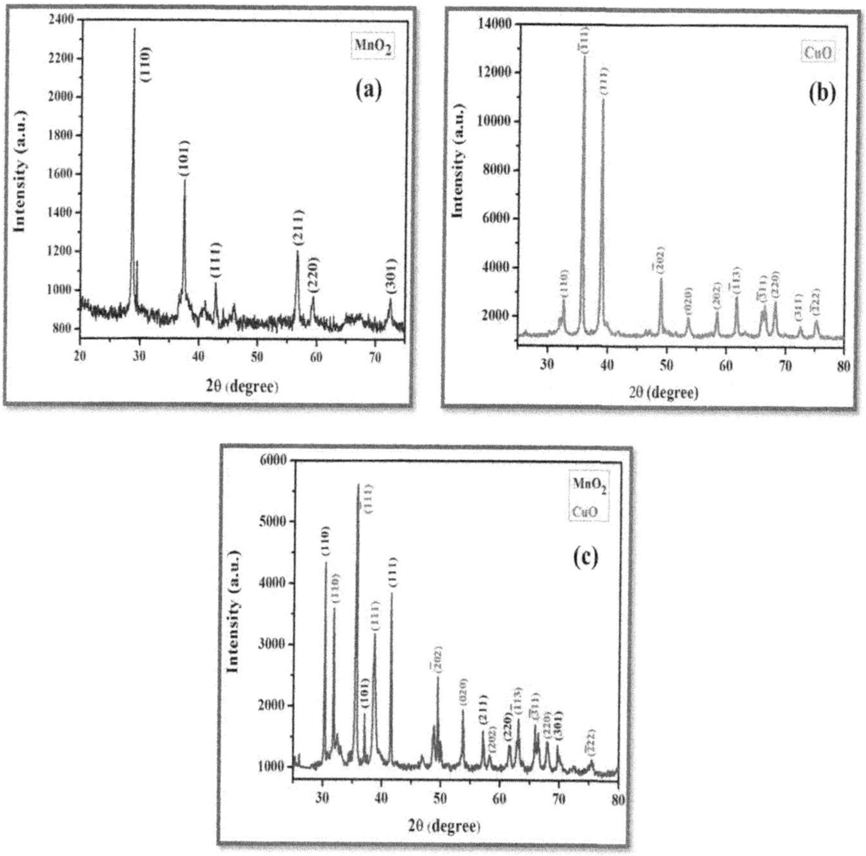

Fig. 3.1: XRD Patterns of (a) MnO_2, (b) CuO and (c) MnO_2/CuO nanocomposite.

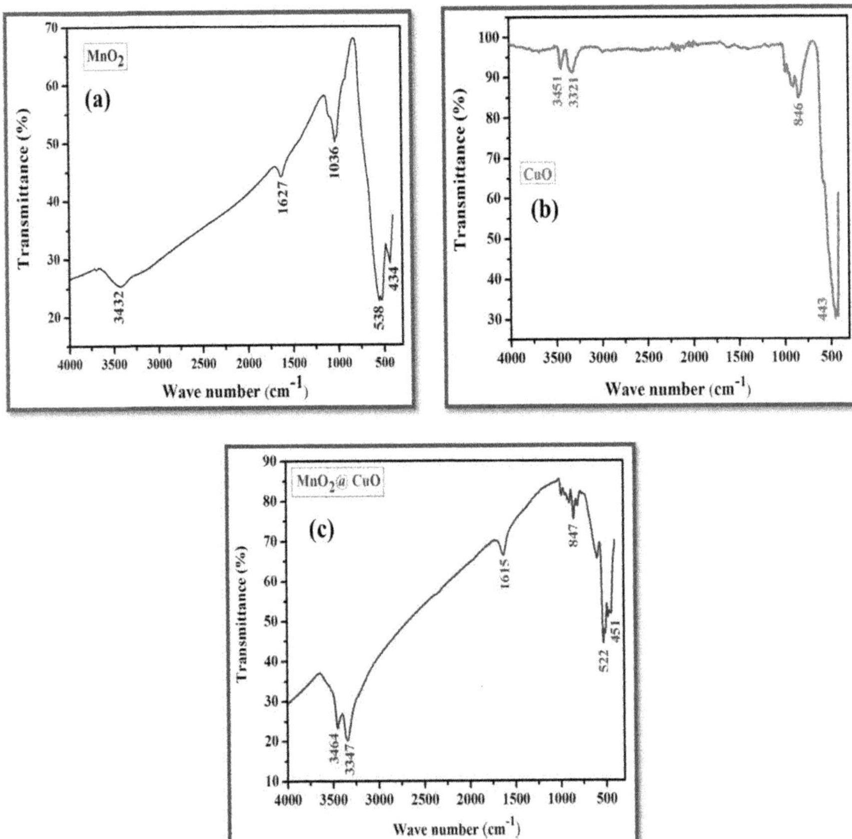

Fig. 3.2: FTIR Spectrum of (a) MnO_2, (b) CuO and (c) MnO_2/CuO nanocomposite.

The peak at 3464 and 3347 cm^{-1} is attributable to stretching vibrations of the O-H adsorbed nanocomposite surfaces MnO_2 and MnO_2/CuO. The peaks at 1615 cm^{-1} were tentatively assigned to the vibration of C–N bond. 1847 cm^{-1} are consigned to bending vibrations of O-H combined with the Cu atoms.

3.3.3 Morphological Studies

The morphologies of selected MnO_2, CuO NPs and their resultant composites of MnO_2/CuO were investigated by HRSEM (Fig. 3.3). Fig. 3.3(a) shows a clear spherical like morphology for MnO_2 samples. The MnO_2 sample exhibits a uniform distribution of spherical nanoparticles with the average size of 20–30 nm. As shown in Fig. 3.3(b), it can be seen that the CuO support nanoparticles are also mainly with the spherical shaped particles. The decoration of CuO NPs into MnO_2 NPs (Fig. 3.3(c)) exhibited the same spherical shaped particles with uniform distribution from the CuO and MnO_2 NPs combination. This sphere like structure may promote an efficient bonding between electrolyte and active electrode, and the distinct morphology of the composite could offer fast transfer channels of ion and electrons. Therefore, it can be expected that MnO_2/CuO nanocomposite provide large specific surface area, and then distributes excellent electrochemical performance.

The conventional EDX (Energy Dispersive Spectroscopy) elemental analysis of MnO_2, CuO NPs and MnO_2/CuO nanocomposite (Fig. 3.4 (a-c)) indicate that manganese, copper and oxygen elements are uniformly distributed in the final composites, which further confirms a successful synthesis of MnO_2/CuO nanocomposite.

The presence of crystalline phases in the composite, the shape and size of the MnO_2, CuO and MnO_2/CuO nanocomposite were studied by TEM images. The selected area electron diffraction (SAED) pattern image provides information about the crystalline nature of the materials. Similar to the HRSEM analyses, the bright-field TEM images (Fig. 3. 5(a-c)) also show that the bare MnO_2, CuO NPs and MnO_2/CuO nanocomposite sample is composed of the less agglomerated spherical nanoparticles. The SAED patterns of all the three samples (Fig. 3.6(a-c)) indicate the MnO_2, CuO NPs and MnO_2/CuO nanocomposites with the existence of different orientations of highly crystalline lattice fringes and, in general, the polycrystalline nature of the nanosphere is also confirmed from the SAED pattern analysis.

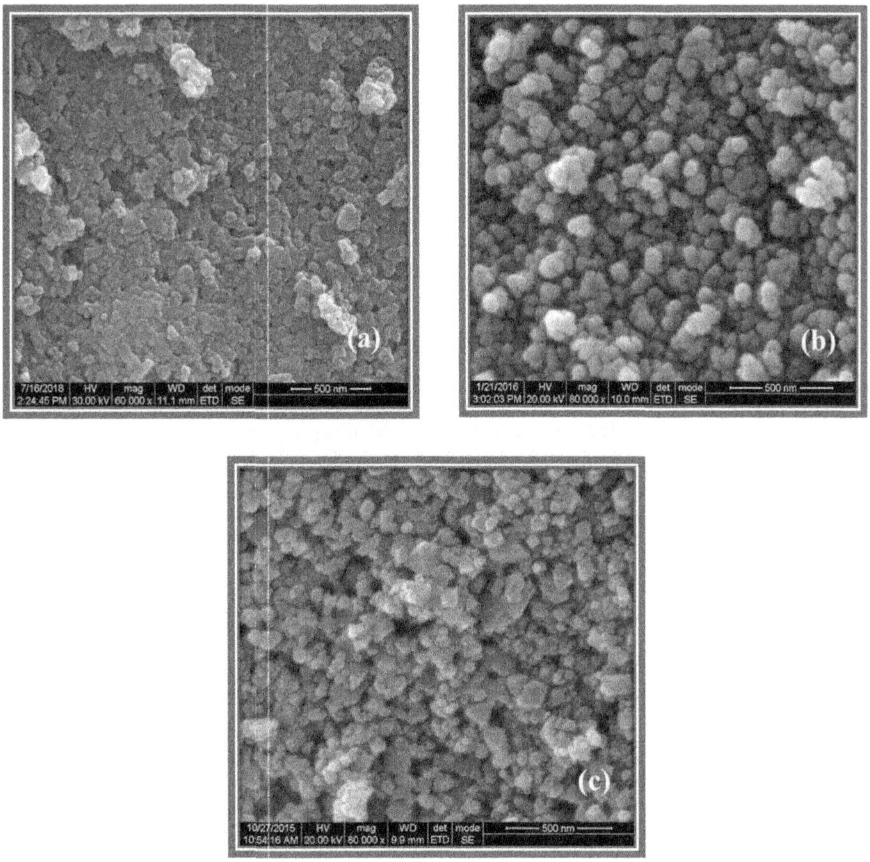

Fig. 3.3: HRSEM images for (a) MnO_2, (b) CuO and (c) MnO_2/CuO nanocomposite.

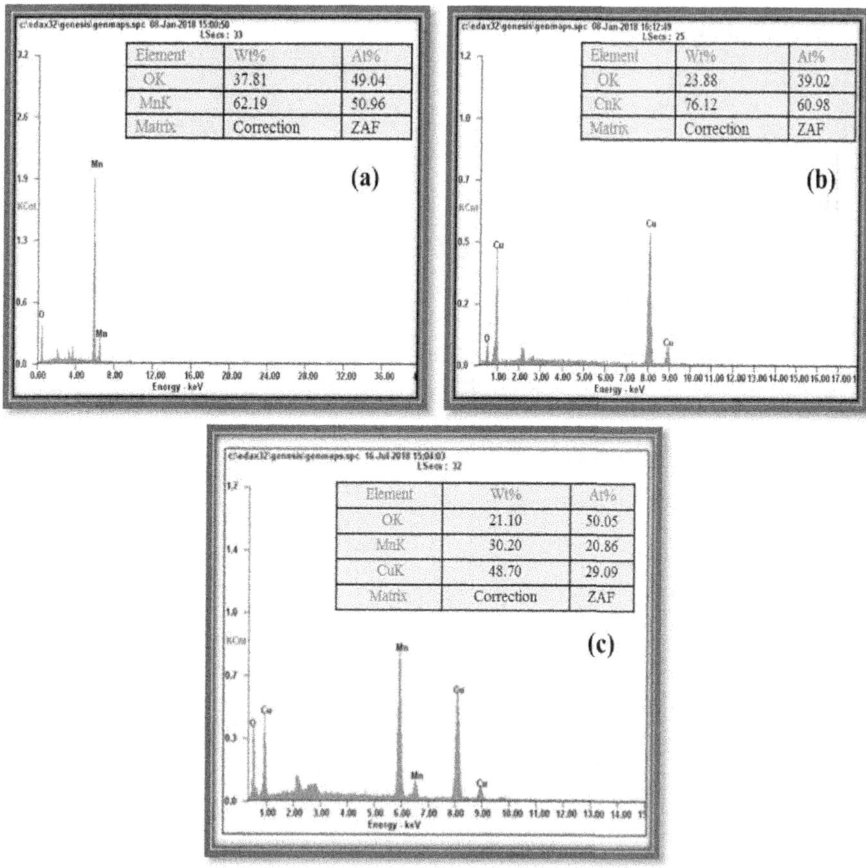

Fig. 3.4: EDX Spectrum of (a) MnO_2, (b) CuO and (c) MnO_2/CuO nanocomposite.

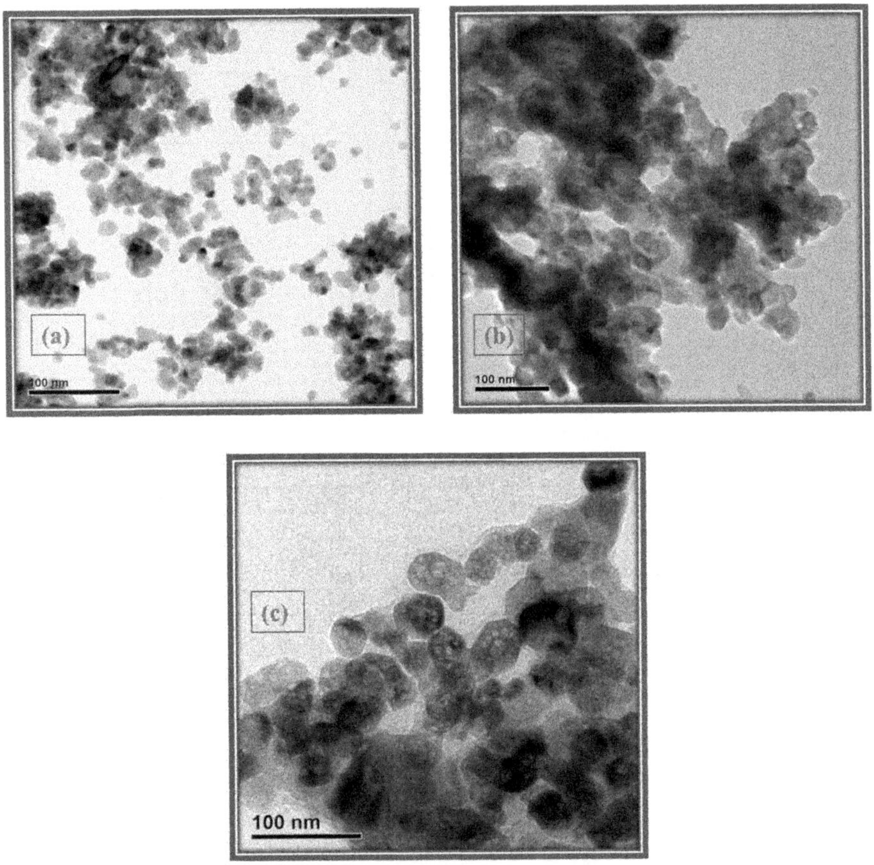

Fig. 3.5: TEM images of (a) MnO_2, (b) CuO and (c) MnO_2/CuO nanocomposite.

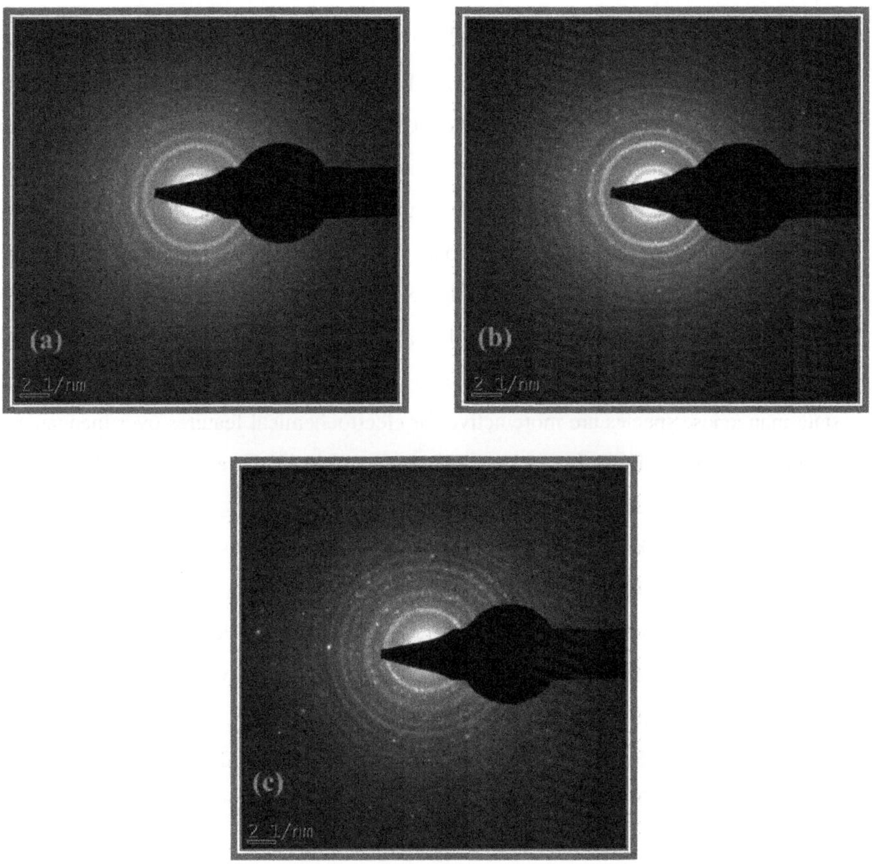

Fig. 3.6: SAED Pattern of (a) MnO_2, (b) CuO and (c) MnO_2/CuO nanocomposite.

3.3.4 XPS Analysis

The chemical surface compositions of MnO_2/CuO nanocomposite were examined using XPS studies. The XPS wide scan data Fig. 3.7 (a) declares the presence of Cu, Mn, and O in prepared composite samples, as well as traces of adventitious C. Fig. 3.7 (b-e) shows the individual scanned spectra of O 1s, C 1s, Mn 2p and Cu 2p elements in the MnO_2/CuO composite sample.

The individual high resolution of Mn 2p for MnO_2/CuO nanocomposite XPS spectrum consists of two prominent peaks at 652.64 eV and 640.91 eV with 11.7 eV of spin energy separation owing to a spin doublet of Mn $2p_{1/2}$ and Mn $2p_{3/2}$ Fig. 3.7(c). Moreover, the atom molar rate of Mn and O is about 1:2 in the survey spectra, demonstrating that the realization of bare MnO_2 and chemical state of manganese element is +4. These results are evidently comparable with the earlier reported data for Mn $2p_{3/2}$ and Mn $2p_{1/2}$ in MnO_2 [Yan. J et al. 2009, Li. Z et al. 2011], further declare that the existence of MnO_2 phase in MnO_2/CuO nanocomposite. The higher oxidation state manganese species are more active for electrochemical features over manganese-based electrodes. Fig. 3.7(d) shows two remarkable peaks around 933.71 eV and 953.68 eV in Cu 2p spectra which are corresponding to Cu $2p_{3/2}$ and Cu $2p_{1/2}$ of copper oxide. The two notable satellite peaks around 941–944 eV and 962.3 eV further declare the existence of CuO [Chen. W et al. 2015, Molazemhoseini. A et al. 2017]. The above outcome specifies that the copper ions are in +2 valence state. The Cu^{2+}presence is endorsed to charge transfer, which results from the interaction between both oxides [Khan. M.A et al. 2018]. The fitted high-resolution O 1s spectrum (Fig. 3.7e) reveals two peaks at 528.32 eV and 530.20 eV binding energy due to oxygen species comprising the MnO_2/CuO nanocomposite while a peak existence of oxygen and H_2O molecules.

3.3.5 Electrochemical Measurements

Cyclic Voltammetry (CV) provides a useful technique in screening potential capacitor materials in a relatively swift procedure. The technique is performed by cycling a linear potential sweep between two defined voltages determined by the electrolyte decomposition limits, and generally aided by a reference electrode in a three-electrode cell.

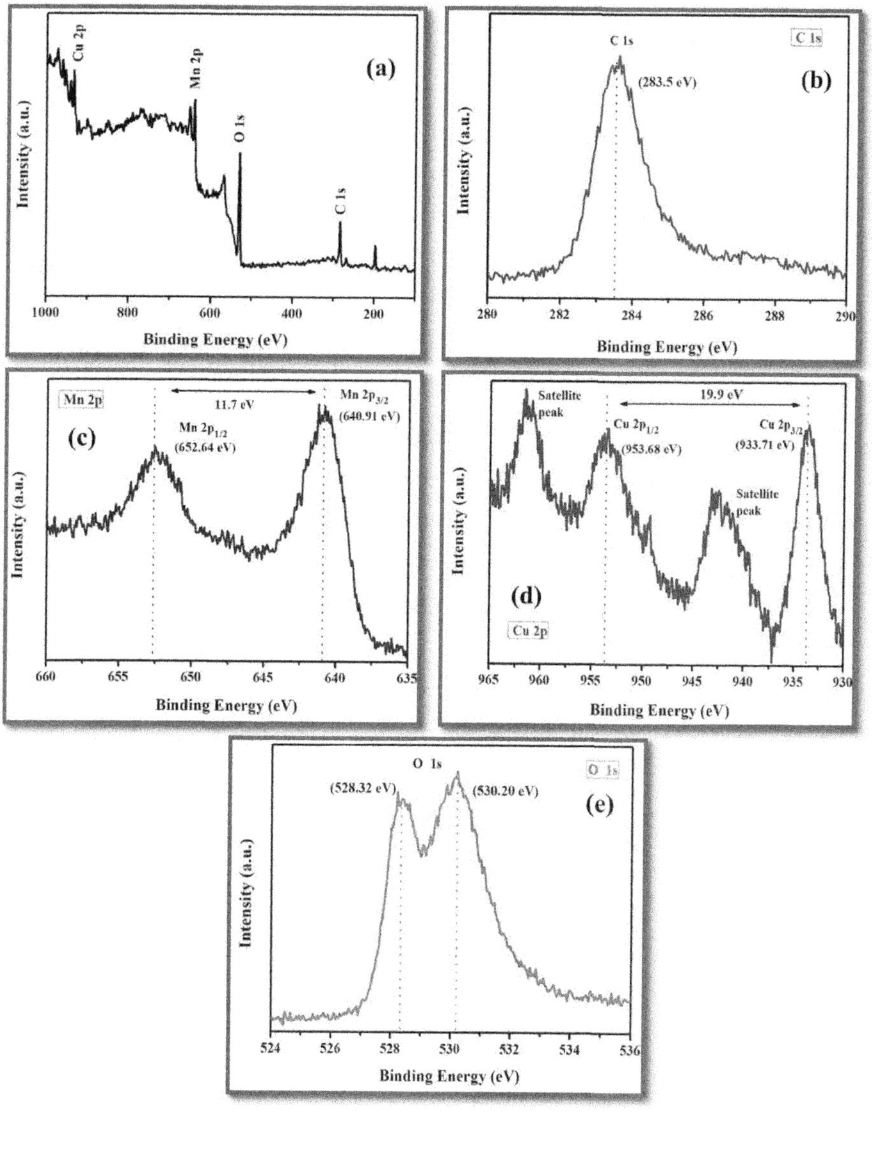

Fig. 3.7: XPS spectra of MnO_2/CuO nanocomposites (a) full XPS survey scan spectrum and (b–e) wide scan XPS spectra of C1s, Mn 2p, Cu 2p and O 1s respectively.

The potential applied to the electrode varies linearly with time, yielding a time-dependent dynamic response current which is recorded by computer. Electrochemical belongings of MnO_2/CuO nanocomposite were examined with a three-electrode system in 1M aqueous KOH electrolyte with Ag/AgCl (reference electrode) and Pt foil (counter electrode). For comparison, electrochemical features of MnO_2 and CuO were also investigated. Fig. 3.8 shows the CV curves of bare MnO_2, CuO NPs and MnO_2/CuO nanocomposite. The CV measurement of an electrode was taken in potential range of -0.1 V to 0.4 V. Fig. 3.8 (a) depicts the bare MnO_2 CV curve, which appears as rectangular shape without noticeable redox peaks; capacitive behavior of MnO_2 nanospheres electrode is pseudocapacitance, attributed to reversible redox reaction among Mn^{3+} and Mn^{4+}.

The shape of the plot is also not fully rectangular, which point out a deviance from ideal behavior of supercapacitor. The change of slight shape in CV curves specifies the excellent charging rate capability of an electrode. Fig. 3.8 (b) illustrates the CV diagrams of pristine CuO active electrode. All CV curves showed a pair of redox peaks at various scan rates from 5–320 mV/s over the voltage range of -0.1–0.4 V, indicating similar pseudocapacitive response and significant rate capability. Additionally, the redox peaks current intensity areas turn into larger as the scan sweep has increased with increasing of scan rates.

Moreover, enormously shifting in anodic peaks towards a positive (+ve) potential owing to electro active sample polarization, while the cathodic peaks transferred towards a negative (–ve) potential with increasing scan rate since the low resistance and fast ion and electron transfer rates. The strong peaks could be closely correlated to the morphology and chemical composition of CuO electrode material for reversible faradaic redox reactions and redox reaction resultant to KOH (Potassium hydroxide) electrolyte. Fig. 3.8 (c) appeared CV plots of MnO_2/CuO nanocomposite at various scan rates from 5–320 mV/s. The CV curves of composite electrode shows relatively apparent redox peaks, which can be ascribed to typical pseudocapacitive performance of MnO_2 electrode.

Bare MnO_2, CuO and MnO_2/CuO nanocomposite electrode was systematically studied by using the charge-discharge curve test in order to measure the capacitance value of the materials and their resultant plots are presented in Fig. 3.9. The GCD test was performed at an increasing current densities of 0.5, 1, 2 and 4 A/g and maintained 0.1 to 0.4 V range during entire GCD test. The specific capacitance values of MnO_2/CuO nanocomposite to be measured by following of the relation:

$$\textbf{Specific Capacitance} = C_{GCD} = \frac{i \times \Delta t}{(\Delta v \times m)} \qquad \textbf{(3.2)}$$

Where *I*, *Δt*, *m* and *ΔV* are discharge current (A), discharging time, mass of an electrode (cm^2), and discharge potential range (V), respectively.

The GCD curves of MnO_2/CuO electrode displays that the curve exhibits equilateral triangle shapes which indicate the excellent reversibility during the entire process.

The specific capacitance for those individual compound of pristine MnO_2 (Fig. 3.9(a)) are calculated as 175.51, 66.78, 45.15, and 37.47 F/g at various current densities of 0.5, 1, 2 and 4 A/g respectively. With an increasing current densities of 0.5, 1, 2 and 4 A/g, CuO specific capacitances values were calculated as 191.06, 97.95, 60.76, 48.13 F/g respectively (Fig. 3.9 (b)) which are greater than some of the earlier reports in literature. The resultant MnO_2/CuO nanocomposite exhibited the specific capacitance values are 279.12, 161.57, 89.48 and 66.21 F/g at current densities of 0.5, 1, 2 and 4 A/g, respectively (Fig. 3.9 (c)). The measured specific capacitance values of MnO_2/CuO (279.12 F/g) electrode are much greater than their individual compound of bare MnO_2 (175.51 F/g) and CuO (191.06 F/g) electrode materials. The improvement in specific capacitance of MnO_2/CuO electrode is due to MnO_2 spherical nanoparticles were well wrapped on the surface of CuO nanoparticles to provide composites with larger surface area. The resultant MnO_2/CuO electrode showed a better specific capacitance than some of MnO_2 decorated CuO composites [Zhang. Z et al. 2015, Zhang. Y et al. 2018, Xin Zhang. Y et al. 2013]. The energy density (E_d) and power density (P_d) of MnO_2/CuO nanocomposites were estimated using equation 3.3 and 3.4 and given in the Table 3.1.

$$\mathbf{E_d = 0.5C_s\Delta V^2} \qquad \textbf{(3.3)}$$

$$\mathbf{P_d = \frac{E_d}{t}} \qquad \textbf{(3.4)}$$

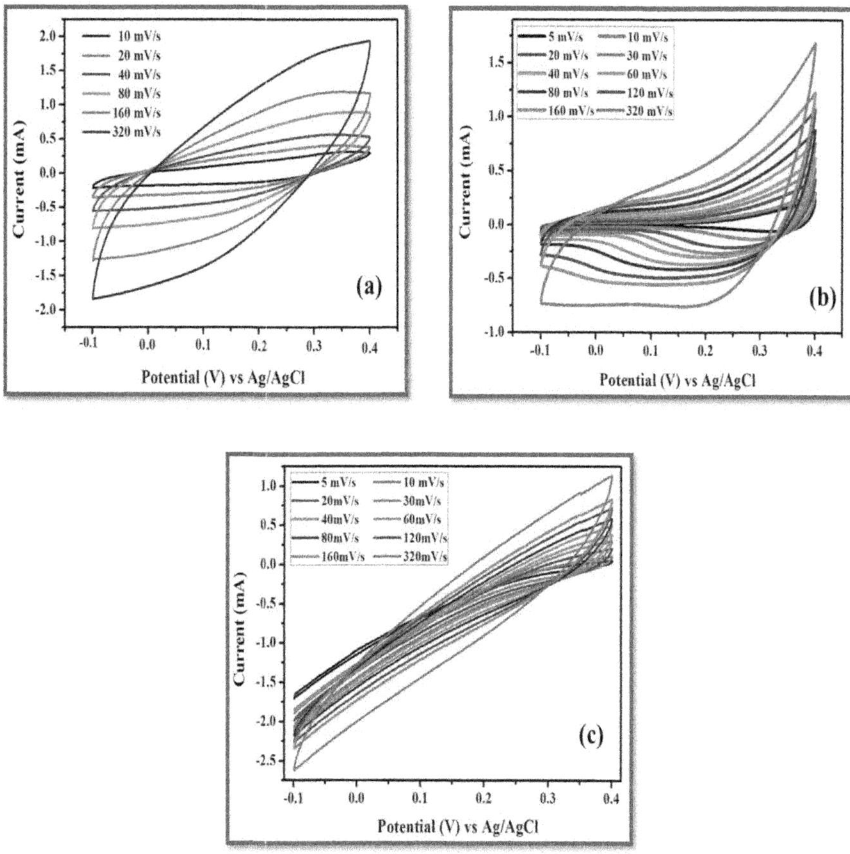

Fig. 3.8: Cyclic Voltammograms of (a) MnO_2, (b) CuO and (c) MnO_2/CuO nanocomposite at different scan rates.

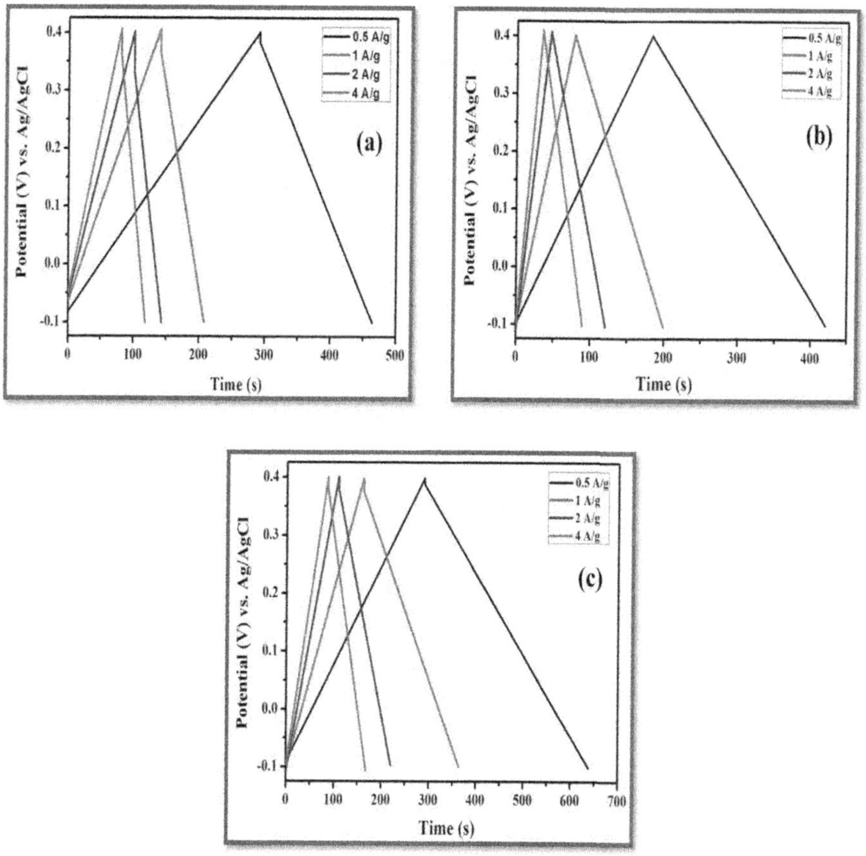

Fig. 3.9: Charge–discharge curves at different current densities of (a) MnO_2, (b) CuO and (c) MnO_2/CuO nanocomposite.

The energy density is found to increase with the power density. The power density is in increasing trend which point out the pseudocapacitor discharging capacity is directly proportional to discharge current. For pure CuO and MnO_2 energy density of 748.9928 Wh kg^{-1} and power density of 0.2500 kW kg^{-1} and energy density of 688.0270 Wh kg^{-1} and power density of 0.1507 kW kg^{-1} were observed at a current density of 0.5 Ag^{-1}. At the same current density of 0.5 Ag^{-1}, the MnO_2/CuO nanocomposites provided a superior energy density of 1094.1504 Wh kg^{-1} and power density of 0.2243 kW kg^{-1}. The calculated measurements of specific capacitance, energy and power density values of pure CuO, MnO_2 and MnO_2/CuO nanocomposites are tabularized in Table 3.1, Table 3.2 summaries the comparisons of specific capacitance performance of MnO_2, CuO, MnO_2/CuO nanocomposites exposed in Fig. 3.10, prepared in the electrode material present work revealed good cycling performance than some of the previously reported papers.

As exposed in Fig. 3.11, the MnO_2/CuO nanosphere-like structure exhibited excellent long-term electrochemical stability and high rate stability with a very slight decrease to 91.26%, even after 10000 cycles. Such exceptional cycling stability of MnO_2/CuO electrode may be endorsed to the following: the sphere like structure, which can improve the volume changes during the charge–discharge process to promise good stability; gradual penetration of the KOH electrolyte ions into an electro-active material; and reduced volume expansion resulting from the rapid and long-term faradaic reaction.

The imaginary component Z'' of the overall impedance Z response relates to the capacitance of the electrode through the equation $Z''=1\ 1/j\omega C$. This relation has placed the ac current impedance technique as a principle method for capacitive behaviour analysis. Impedance analysis using ac provides a means of measuring capacitance as a function of the frequency ω. In addition, evaluation of the phase relation between Z and Z'' impedances generated by this procedure can be achieved. Important circuit elements of the electrode structure or constructed device, including the ESR, and potential-dependent Faradaic resistances (e.g., pseudocapacitive, leakage resistance) can be separated and determined using this technique. The examination of fundamental and applied characteristics of ECs can be performed using impedance spectroscopy over a wide range of frequencies (10 kHz to 10 MHz).

Table 3.1: Specific Capacitance, Energy and Power density values of MnO_2, CuO and MnO_2/CuO Nanocomposite

Sample	Current density (Ag^{-1})	Specific Capacitance (Fg^{-1})	Energy density ($Wh \cdot kg^{-1}$)	Power density ($kW \cdot kg^{-1}$)
MnO_2	0.5	175.51	688.0270	0.1507
	1	66.78	261.7838	0.3015
	2	45.15	176.6254	0.6031
	4	37.47	146.8871	1.2062
CuO	0.5	191.06	748.9928	0.2500
	1	97.95	383.9875	0.5001
	2	60.76	238.1886	1.0006
	4	48.13	188.6903	2.0003
MnO_2/CuO	0.5	279.12	1094.1504	0.2243
	1	161.57	633.3606	0.4487
	2	89.48	350.7616	0.8974
	4	66.21	259.5306	1.7948

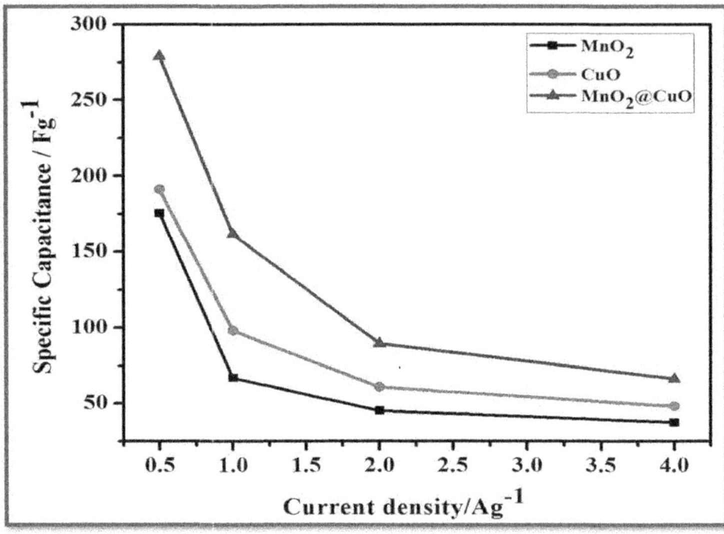

Fig. 3.10: Specific capacitance (Cs) versus discharge current density of MnO_2, CuO and MnO_2/CuO nanocomposite electrodes.

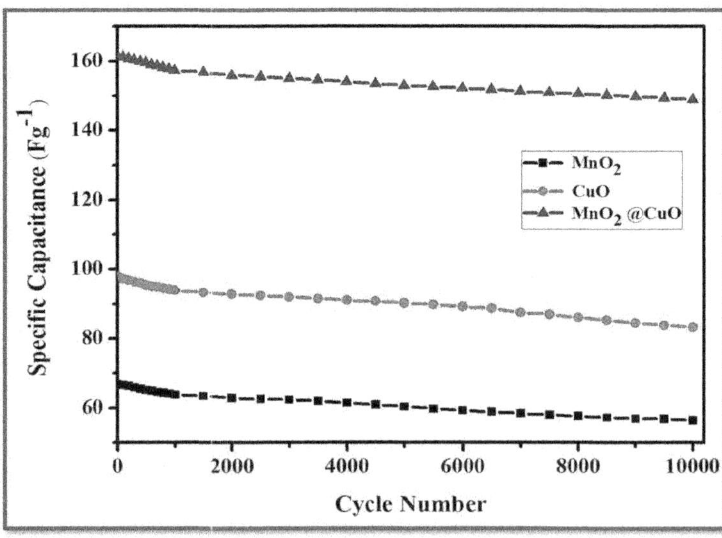

Fig. 3.11: Cycling performance of the MnO_2, CuO and MnO_2/CuO nanocomposite electrodes.

Table 3.2: Comparative Studies of MnO_2 or CuO based Composites of Supercapacitor with Various Methods and Nanostructure

Compounds	Nanostructures	Electrolyte	Specific Capacitance	Reference
CuO/MnO_2	Spherical	1 M Na_2SO_4	228 F/g	[Zhang. Z et al. 2015]
CuO/MnO_2	Spherical	1 M Na_2SO_4	240 F/g at 0.5 A/g	[Zhang. Y et al. 2018]
MnO_2/CuO	Nanosheets/ flower	1 M Na_2SO_4	167.2 F/g at 0.3 A/g	[Xin Zhang. Y et al. 2013]
$CuO@MnO_2$ Core-shell	Nanosheets	1 M Na_2SO_4	252.6 F/g at 0.1 A/g	[Guo. X-L et al. 2015]
MnO_2/NiO	Spherical	1 M KOH	247 F/g at 0.5 A/g	[Mohamed Racik. K et al. 2019]
MnO_2-based mixed oxides	Amorphous	1M Na_2SO_4	210 F/g at (0.12 A/g)	[Kim. H and Popov. B.N 2003]
Nanosized Ni–Mn oxides	Spherical	4 M KOH	198 F/g	[Zhou. J et al. 2006]
CuO/Co_2O_4	Nanosheets	5M KOH	246 F/g	[Ramesh. S et al. 2019]
CuO/PANI	Spherical	1M Na_2SO_4	185 F/g at 5 mV/ s	[Gholivand. M. B et al. 2015]
MnO_2/CuO	**Spherical**	**1M KOH**	**279.12 F/g at 0.5 A/g**	**Present work**

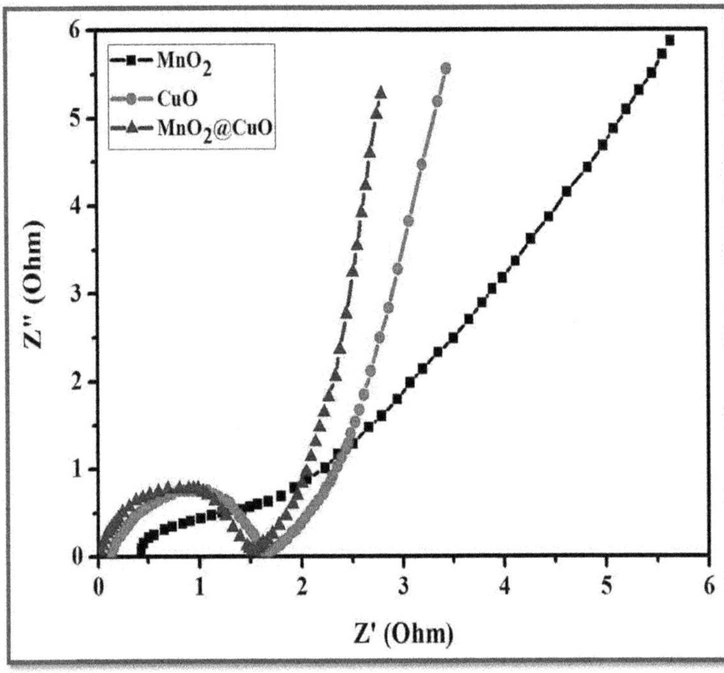

Fig. 3.12: Nyquist plots of MnO_2, CuO and MnO_2/CuO nanocomposite electrodes.

The impedance spectrum studies of bare CuO, MnO_2 and MnO_2/CuO nanocomposite electrodes were studied using EIS analysis. Fig. 3.12 shows the Nyquist plots of all the samples: it shows a semicircle at high frequencies and a straight line at low frequencies. An almost vertical line represents quick ion diffusion in the electrolyte and represents model capacitive performance in the charge–discharge process. The semicircle at the higher frequency region could be characteristics of the resistance of oxide-electrolyte interface. The resistance contribution from the faradaic redox procedure is involved in this region. The lower frequency region can be the characteristics of Warburg impedance, which defines diffusive resistance of the OH- ions inside the electrode material. The plot displays the linear line along the y axis, which is attributed to the Faradaic behavior owing to the reversible Faradaic redox reaction. The entire experimental results of MnO_2/CuO nanocomposite declared that the composite electrode based on incorporation of CuO into MnO_2 is suitable electrode candidate material for supercapacitors.

3. 4. CONCLUSION

Metal oxide composites can combine the merit of individual, and synergetic effects can provide more structures, morphologies, and properties than single ingredient. In addition, MnO_2 composites can heighten electrical conductivity and provide more reaction active sites. The structural and morphological characterization methods are discussed in detail. The structural characteristics of the electrode materials were carried out by using X-ray diffraction (XRD), Fourier transform infrared (FT-IR) and X-ray photoelectron spectroscopy (XPS). The shape and size of the materials were studied by collecting scanning electron microscopy (SEM), transmission electron microscopy (TEM) images. The electrical and electrochemical properties were studied by cyclic voltammetry (CV), galvanostatic charge-discharge (GCD) and electrochemical impedance spectroscopy (EIS) analysis. The specific capacitance, energy and power density of the electrode and supercapacitor device were calculated from the CD plot. The electrochemical performances of economical metal oxides such as CuO and MnO_2 and MnO_2/CuO nanocomposite using hydrothermal method are elaborately discussed in this chapter. The main advantages of the wet chemical method are its simple and the level of experimental control. However, based on developed knowledge, the supercapacitance properties of the MnO_2/CuO nanocomposites have attracted great attentions. It is found that the obtained MnO_2/CuO nanocomposites exhibit excellent capacitance of 279.12 Fg^{-1} at a current density of 0.5 Ag^{-1}, and the capacitance retention approaches 91.26% after 10000 cycles, thus showing outstanding cycling stability. When tested as supercapacitor electrodes in aqueous electrolyte, the nanocomposites of MnO_2/CuO exhibited enhanced specific capacitance and cyclic stability compared to the bare MnO_2 and CuO nanoparticles. The improved electrochemical performances of the MnO_2/CuO nanocomposites could be attributed to the synergistic effects between the MnO_2 and CuO nanostructures. These results are encouraging from the view point that low cost materials prepared by environmental benign process could be used as promising electrode materials for energy storage devices.

CHAPTER IV

SYNTHESIS AND CHARACTERIZATION OF $MnO_2/\alpha\text{-}Fe_2O_3$ NANOCOMPOSITE FOR SUPERCAPACITOR APPLICATION

4.1. INTRODUCTION

The exponential growth of economy demands, the supply of energy devices to meet the global standards. The efficient and clean energy devices are desired due to energy and environment based crisis [Seevakan. K et al. 2018, Huanhuan Wang et al. 2016, Seevakan. K et al. 2018]. Nowadays, tremendous efforts and research were carried out to invent energy storage and conversion devices from sustainable and renewable clean energy sources because of the imminent shortage of fossil fuels and growing environmental concerns [Xu. X et al. 2014]. Supercapacitors or Ultracapacitors is a significant electrochemical energy storage system catching more and more attention in current years which is endorsed to their high power density, good stability, low cost, long cycle life, environmental friendliness and other excellent properties [Shuijian. H et al 2015]. The gap among batteries, conventional solid state and electrolytic capacitors are appropriately filled by supercapacitors and they can play an important role in supplementing or replacing batteries [Zhenyuan. J et al. 2014].

Normally electrode materials were divided into two categories based on charge storage mechanism: (i) electric double-layer capacitors (EDLCs) and (ii) pseudocapacitors (PCs) [Zheng. Y et al. 2016, Bhattacharjya. D et al. 2013]. In general, EDLCs contains conductive porous carbon materials, such as graphene, super carbon, carbon spheres and activated carbon [Yin. J et al. 2014, Wei. W et al. 2014]. The higher specific capacitances and energy densities in prime pseudocapacitive materials such as some metal oxides, conducting polymers and hydroxide particles are due to fast, reversible, multi-electron, surface faradic reactions [Mi. L et al. 2015, Wei. W et al. 2017, Hsu. L.C et al. 2011]. In specific, there are certain metal oxides and low-price hydroxides, such as ZnO, $\alpha\text{-}Fe_2O_3$, MnO_2, Mn_3O_4, Co_3O_4, NiO and $Co(OH)_2$ were reported as electrode candidates.

Hematite (α-Fe_2O_3), an important n-type semiconductor with a hexagonal corundum structure, is regarded as the most stable iron oxide, which has promising applications in photoelectronics, catalysis, gas sensing, field emission ,energy storage, and sewage treatment, due to its superior environmental compatibility, abundant resource and higher corrosion resistance [Zboril. R et al. 2002]. Generally α-Fe_2O_3 shows fascinating magnetic features, that is to say, only weak or parasitic ferromagnetism between Morin and Neel temperature [Chou. S.L et al. 2010]. Furthermore, even though most transition metal oxides, when engaged as energetic electrode materials for supercapacitor electrodes, exhibit excellent pseudocapacitive behaviors with long-standing cycling stability, the experimentally observed specific capacitance of α-Fe_2O_3 is still far below the theoretical value, due to its poor electrical conductivity [Ready. D. W 1966]. Fe_2O_3 being one of the most abundant material is significantly advantageous due to low cost, environmental friendliness and from safety point of view. However, their intrinsic poor conductivity and aggregated morphology restricts their extensive use as electrode material. Tang and his co-workers (2015) developed Fe_2O_3 nanowires (NW) with 10 nm using modified hydrothermal and post calcination process. The Fe_2O_3 NWs exhibited specific capacitance of 908 Fg^{-1} at 2 Ag^{-1} in a three-electrode system and showed 90% retention in capacitance up to 10 Ag^{-1}. [Chaudhari et al. 2013] have synthesized 1D porous α-Fe_2O_3 nanorods via a wet chemical method without using any template. The nanorods delivered a specific capacitance of 308 F g^{-1} at a current density of 0.1 Ag^{-1} in 1 M H_3PO_4 electrolyte. The excellent electrochemical performance is attributed to the high surface area induced by the 1D porous nanorod structures. [Shivakumara et al. 2014] prepared porous α- Fe_2O_3 nanostructures via sol-gel route, and reported a specific capacitance of 193 Fg^{-1} at a current density of 1 Ag^{-1} in 0.5 M Na_2SO_4 solution. The high capacitance and stability of α- Fe_2O_3 electrode is attributed to large surface area and porosity of the material. [Rongmei Liu et al. 2015] synthesized novel red blood cell shape α-Fe_2O_3 microstructure and FeO(OH) nanorods through hydrothermal method using ammonium chloride as structure directing agent. The FeO(OH) nanorods exhibited higher capacitance (392.6 Fg^{-1}) than red blood cell shape α-Fe_2O_3 microstructure (118.1 Fg^{-1}) at a current density of 1 Ag^{-1}. This high performance was attributed to the desirable morphology and component.

Manganese oxides have several different forms and phases (such as MnO_2, Mn_2O_3, MnO and Mn_3O_4) MnO_2 is one of the most fascinating manganese oxides for Electrochemical Energy Storage (EES) applications due to its high capacity/capacitance, low cost and environmental friendliness. A great many MnO_2 based nano and microstructures have been reported and fabricated by different solution-based synthesis methods, and enhanced electrochemical performances have been established in these systems with promising practical applications. HT is one of the three most popular solution-based methods (hydrothermal synthesis, chemical precipitation and electro-deposition methods) for preparation of MnO_2 and MnO_2 based composites for EES. According to the reaction mechanisms and reagents, basically, there are about three kinds of HT methods for the fabrication of MnO_2 and MnO_2 based composites for EES. The manganese oxide (MnO_2) has high value of theoretical capacitance 1100-1300 Fg^{-1} [Huang. Y et al. 2013]. Also it is proved to achieve high experimental capacitance (600-800 Fg^{-1}) and better cycling stability up to thousands of cycles [Kai. K et al. 2012, Sarkar. D et al. 2013]. Recently few admirable mixed oxide nanostructures for high specific capacitance and stability were reported. For instance, Sarkar et al. (2013) reported high performance capacitor electrodes based on α-Fe_2O_3/MnO_2 core–shell nanowire hetero structure (NHs) arrays. Nearly 801 Fg^{-1} of specific capacitance have been observed for the α-Fe_2O_3/MnO_2 NHs electrode at a current density of 1 Ag^{-1}. Nie. G and Co-Workers (2017) reported core-shell typed heterostructures of α-Fe_2O_3@MnO_2, the composite contains 60.1 wt% of MnO_2, provides specific capacitance with higher value of 289 Fg^{-1} at 1.0 Ag^{-1}, a good rate capability of 40% at 5Ag^{-1} and a greater cycling stability of 85% up to 1200 cycles than the pristine MnO_2, emphasizing the benefits of such unique configuration combined by synergistic effect. The previous supercapacitor literature shows that minimum amounts of additives can dramatically influence the magnetic and opto-electronic properties of ferrimagnets by changing the nature and concentration defects [Liu. J. P et al. 2012]. Therefore, it was possible for hybrid metal oxides, such as α-Fe_2O_3–MnO_2, MnO_2–NiO and Co_3O_4–MnO_2 to achieve enriched electrochemical properties by combining preferred functions of individual components and intruding synergistic effects into the pseudocapacitor electrode system [Wang. G et al. 2012]. [Nie. G et al. 2017] reported core-shell typed heterostructures of α-Fe_2O_3@MnO_2, the composite contains 60.1 wt%

of MnO_2, provides specific capacitance with higher value of 289 Fg^{-1} at 1.0 Ag^{-1}, a good rate capability of 40% at 5Ag^{-1} and a greater cycling stability of 85% up to 1200 cycles than the pristine MnO_2, emphasizing the benefits of such unique configuration combined by synergistic effect. Zhu. L and Co-Workers (2015) reported the hydrothermal preparation of core-shell $MnO_2@Fe_2O_3$ nanospindle, which exhibits the specific capacitance of 159 Fg^{-1} at 0.1 Ag^{-1} current density and specifically 97% capacitance retention after 5000 cycles in a 0.5 mol/L Potassium sulphate (K_2SO_4) electrolyte. In the view from above considerations, herein manganese dioxide/ iron oxide (MnO_2/ α-Fe_2O_3) spherical nanocomposite was successfully prepared by facile hydrothermal route. The structure and optical morphology of bare MnO_2, α-Fe_2O_3 and MnO_2/α-Fe_2O_3 nanocomposite were examined by XRD, HRSEM, TEM, FT-IR, VSM and XPS techniques. The electrochemical studies were inspected by CV, GCD and EIS techniques. The results emphasize that the MnO_2/α-Fe_2O_3 nanocomposites electrodes exhibit high rate capability, high energy density, high power density and also long cycle stability, which make this type of environment friendly, nanostructures a perfectly promising material for fabrication of supercapacitors electrode.

4. 2 EXPERIMENTAL

4.2.1 Preparation of MnO_2 Nanoparticles

For the hydrothermal synthesis of MnO_2 nanoparticles, the stoichiometric amount of $MnCl_2 \cdot 4H_2O$ and urea were dissolved in 50 mL of distilled water to form a homogeneous solution. Then the mixed solution from above was transferred into Teflon autoclave and set aside at 120° C for 24 hours. The resulting powders were centrifuged using with DI water and methanol. Finally, the precipitate was dried at 80° C for 12h. Finally, the MnO_2 powder was calcinated at 400° C for 3 h.

4.2.2 Preparation of α-Fe_2O_3 Nanoparticles

For the hydrothermal synthesis of Fe_2O_3 nanoparticles, 1M of ferric chloride ($FeCl_3.6H_2O$) was dissolved in 40 mL of DI water. Then mixed solution of ferric chloride and deionized water was transferred into a 100 mL capable Teflon-lined autoclave filled with deionized water till up to 80% of the total volume, and the container was maintained at 220° C for 5 h. The settled precipitate was then washed several times using deionized water and methanol. Then the precipitated product was dried at 60° C for 24h and calcinated at 600° C for 2h.

4.2.3 Preparation of MnO_2/α-Fe_2O_3 Nanocomposite

For the MnO_2/α-Fe_2O_3 nanocomposite preparation, the above synthesized α-Fe_2O_3 was dissolved in 20 mL double distilled water by magnetic stirring, then $MnCl_2 \cdot 4H_2O$ and urea was mixed into the dispersion. After constantly stirring for 1h, the mixed solution was transferred into a 100 mL autoclave treated at 120 °C for 24 h. The settled precipitate for α- MnO_2/α-Fe_2O_3 nanocomposite powders was centrifuged using with distilled water and methanol. The resulting product was dried at 80° C for 12 h and annealed at 400° C for 4h.

4.2.4 Characterizations

The crystallographic nature of synthesized pristine powders and the nanocomposites was confirmed by powder XRD analysis using Rigaku D/max-kA diffractometer with CuKα (λ=1.54056 nm).The FT-IR study was performed in the frequency range from 4000-400 cm^{-1} (Perkin-Elmer spectrometer) to confirm the functional groups and co-ordination. The morphologies and elemental composition were observed by scanning electron microscopy (HRSEM, FEI Quanta FEG 200), which is associated with EDX and the TEM images were obtained on a Philips CM200FEG field emission microscope. The room temperature magnetization measurements were performed using a vibrating sample magnetometer (VSM, Lake Shore VSM-7410) with a maximum applied magnetic field of 15 kOe. The surface electronic states presence was analysed by X-ray photoelectron spectroscopy (XPS, Perkin-Elmer PHI 5000 Versa Probe III).

4.2.5 Electrochemical Measurements

Electrochemical properties of bare MnO_2, α-Fe_2O_3 and MnO_2/α-Fe_2O_3 nanocomposite were investigated by using cyclic voltammetry (CV), Galvanostatic Charge and Discharge (GCD) and Electrochemical Impedance Spectroscopy (EIS). All the electrochemical experimentations were executed at room temperature. Cyclic voltammograms of the electrodes were recorded within a potential range between -0.1 to 0.4 V (vs. Ag/AgCl) in different scan rates from 5 to 320 mV/s. The specific capacitance of the electrodes was estimated over the charge/discharge curves as of the

CP and CV measurements. The charge/discharge experiments were performed using chronopotentiometry by applying different constant current range between -0.1 and 0.4 V (vs. Ag/AgCl). The electrochemical impedance was also studied in frequency region between 1Hz and 1MHz. The working electrode preparation consists of active material (80% by weight), activated carbon (10%) and polyvinylidene fluoride (PVDF) suspension (10%) ratio using the solvent of N-methyl-2-pyrrolidone to make homogeneous slurry [Mohamed Racik. K et al. 2019]. Then the slurry from the mixture was pasted on the surface of nickel foil sheet current collector and dried out at 80° C for 24h. All measurements were taken using an aqueous 1M KOH as electrolyte at ambient conditions with the electrode material prepared as a working electrode, a platinum wire as a counter electrode and Ag/AgCl as a reference electrode.

4.3 RESULTS AND DISCUSSIONS

4.3.1 Powder XRD Analysis

XRD patterns have been used to examine the crystalline nature and the phases of the synthesized MnO_2, α-Fe_2O_3 nanoparticles and MnO_2/α-Fe_2O_3 nanocomposite. Fig. 4.1 shows the XRD patterns of MnO_2, α-$Fe_2O_3$and MnO_2/α-Fe_2O_3 nanocomposite. For MnO_2,the peaks appear at 2θ of 28.6°, 37.4°, 42.8°,56.7°,59.3°, and 72.4° can be indexed to be (110), (101), (111), (211), (220) and (301) crystal planes of tetragonal MnO_2 (JCPDS card no. 81-2261) [H. Wang et al. 2017]. No further peaks were detected, specifying the high degree of phase purity of prepared MnO_2 in this work. The prominent peaks appear at 2θ of 24.18°, 33.20°, 35.68°, 40.92°, 49.52°, 54.13°, 57.61°, 62.50°, 64.06°, 72.04° and 75.58°can be indexed to be (012), (104), (110), (113), (024), (116), (018), (214), (300), (119) and (220) which consists of peaks that can be perfectly indexed with the characteristic diffraction peaks of hexagonal structured α-Fe_2O_3 (JCPDS card no. 33-0664) [Qin. W et al. 2011]. They agree with the formation of phase pure hematite (α-Fe_2O_3) structures without any impurity traces. XRD result confirms the successful preparation of MnO_2/α-Fe_2O_3 composites. The average crystallite size of α-Fe_2O_3 and MnO_2 is calculated to be 33.89 nm, 35.89 nm respectively. All the diffraction peaks of MnO_2/α-Fe_2O_3 nanocomposite are established to the MnO_2 with cubic phase and other peaks are corresponding due to α-Fe_2O_3. The XRD results point out the coexistence of $MnO_2$and α-Fe_2O_3.

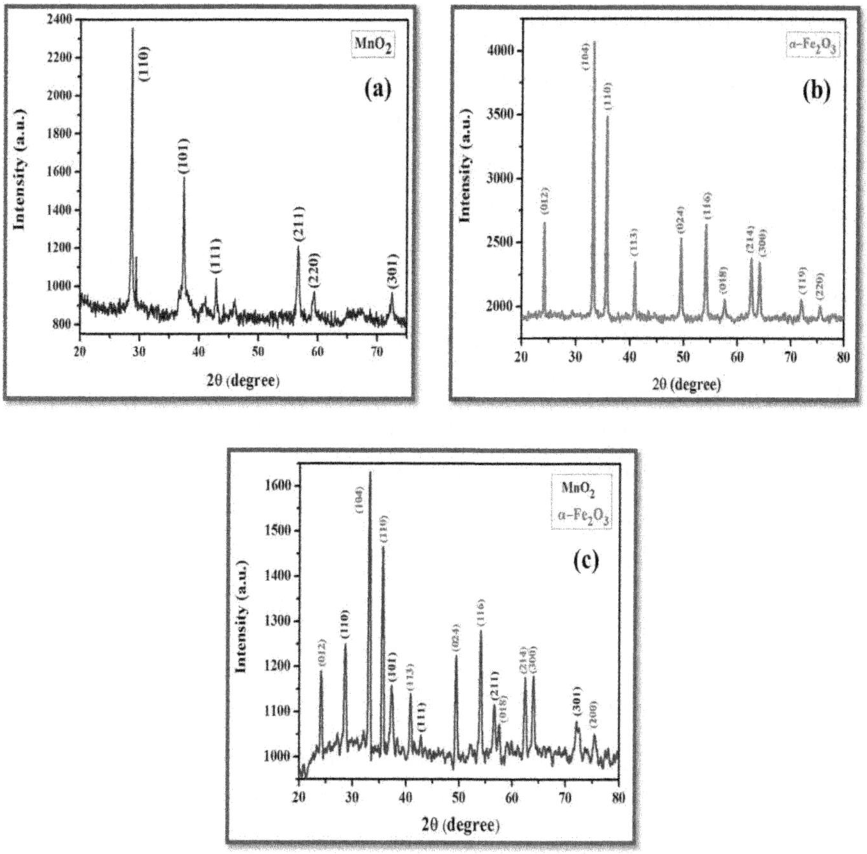

Fig. 4.1: XRD Patterns of Pure MnO_2, α-Fe_2O_3 and MnO_2/α-Fe_2O_3 Nanocomposite.

4.3.2 FTIR Analysis

The vibrational atom modes of MnO_2, α-Fe_2O_3, and MnO_2/α-Fe_2O_3 nanocomposites were confirmed by FTIR analysis as shown in Fig. 4.2 from MnO_2 spectra, the wide band around 3432 cm^{-1} is attributed to O-H stretching vibrations, indicating the hydrates nature of MnO_2. The absorption bands around 1627 and 1036 cm^{-1} are indexed to bending vibrations of O-H combined with the Mn atoms. The peaks located around 538 and 434 cm^{-1} are ascribed to Mn-O vibrations. From α-Fe_2O_3 peaks at 3515 cm^{-1} are assigned as O–H stretching vibration. The vibrations bands at 540 and 466 cm^{-1} attributed as Fe-O bond vibration of Fe_2O_3 [Nag. S et al. 2016]. The formation of MnO_2/α-Fe_2O_3 nanocomposite is confirmed with two strong bands at 538 and 470 cm^{-1}. These bands indicated the division of the vibratory band corresponding to the Fe-O or O-Mn-O bond [Ma. M et al. 2003]. The peak at 3446 cm^{-1} is attributable to stretching vibrations of the O-H adsorbed nanocomposite surfaces MnO_2 and MnO_2/α-Fe_2O_3. The peaks at around 1645 cm^{-1} were tentatively assigned to the vibration of C–N bond. 1036 cm^{-1} are consigned to bending vibrations of O-H combined with the Mn atoms. Thus, the FT-IR spectrum of nanocomposite powders confirms the incorporation of α-Fe_2O_3 nanoparticles in MnO_2. All the typical peaks perceived from the spectra also confirm the formation of MnO_2/α-Fe_2O_3 nanocomposite.

4.3.3 Morphological Studies

Fig. 4.3 displays the HRSEM images of MnO_2, α-Fe_2O_3 and MnO_2/α-Fe_2O_3 nanocomposite and confirmed that all the synthesized samples having the size in non-uniform. Hence it may be the outcome of the combination of the spherical-shaped particles and the growth of non-equal crystalline grains during the preparation period. Due to low dosage of urea to precursor ratio of 1:1, the number of urea around Fe and Mn was very limited in the solution. In the course of the reaction, as the small amount of urea could not provide sufficient hydroxyl groups, therefore non-uniformly sized spherical like particles was formed, and the shape varied from spherical to rod like morphology after the reaction was carried out for 24 h. Similar results were found earlier for CeO_2-ZrO_2 nanocomposites [Zhang. X et al. 2015].

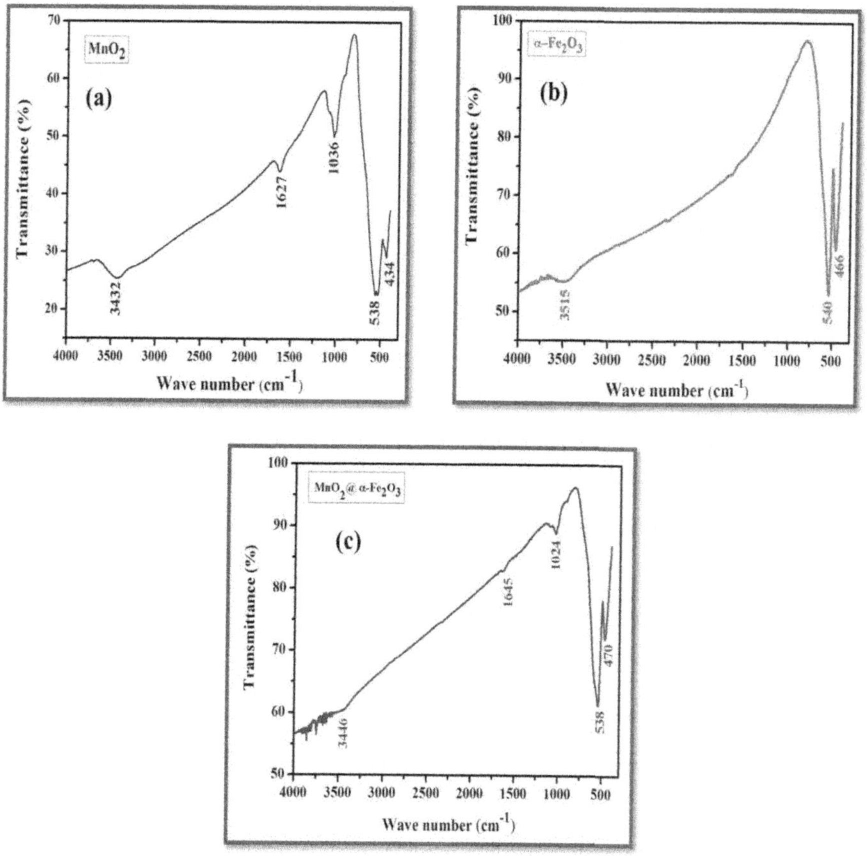

Fig. 4.2: FTIR Spectrum of MnO_2, α-Fe_2O_3 and MnO_2/α-Fe_2O_3 Nanocomposite.

From the HRSEM images, we concluded that the formation of spherical nanoparticles like structure, due to aggregation followed by the agglomeration of nanoparticles. It was illustrious that the particles are collected together, due to magnetic interaction between the particles [Sonia. M.M.L et al. 2018, Amaliya. A.P et al. 2018]. The elemental composition analysis for an active electrode material is essential. The elemental or chemical compositions of MnO_2, α-Fe_2O_3 and MnO_2/α-Fe_2O_3 nanocomposite were studied by EDX analysis. The EDX spectra of prepared samples were shown in Fig. 4.4 (a) & (b) reveal the presence of stoichiometric proportion of O, Mn and O, Fe in MnO_2 and α-Fe_2O_3 samples respectively and Fig. 4.4(c) evidently confirms the presence of Mn, Fe, and O in MnO_2/α-Fe_2O_3 nanocomposite. The observed molar ratio is confirming the formation of MnO_2, α-Fe_2O_3 and MnO_2/α-Fe_2O_3 nanocomposites (Fig. 4.4 (a-c) inset)

The morphology and structural features of MnO_2, α-Fe_2O_3 and MnO_2/α-Fe_2O_3 nanocomposite were described by HR-TEM analysis. The characteristic HR-TEM optical images are exposed in Fig. 4.5 (a-c). It also clearly confirmed the formation of spherical particles with average particle diameter in the range of 25-50 nm from highly magnified TEM image (Fig. 4.5). The SAED pattern of pristine MnO_2, α-Fe_2O_3 and MnO_2/α-Fe_2O_3 nanocomposite were shown in Fig. 4.6(a-c) confirmed the powders were well crystallized in nature.

4.3.4 Magnetic Studies

Using a vibrating sample magnetometer, magnetic measurements of magnetization (M) versus applied field (H) of as-synthesized (a) MnO_2 (b) α-Fe_2O_3, and (c) MnO_2/α-Fe_2O_3 nanocomposite were carried out at room temperature and the hysteresis loops are presented in Figs. 4.7(a-c). From the VSM hysteresis loop graphs, the magnetic characteristic parameters includes coercivity (Hc), saturation magnetization (Ms) and remanent magnetization (Mr) were calculated. The saturation magnetizations (Ms), remanent magnetization (Mr) and coercivity (Hc) were found to be 0.446 emu/g, 0.1452 emu/g and 3777.9 Oe respectively for hematite nanoparticles from Fig. 4.7(b). These results reveal the existence of ferromagnetism behaviour in hematite nanoparticles.

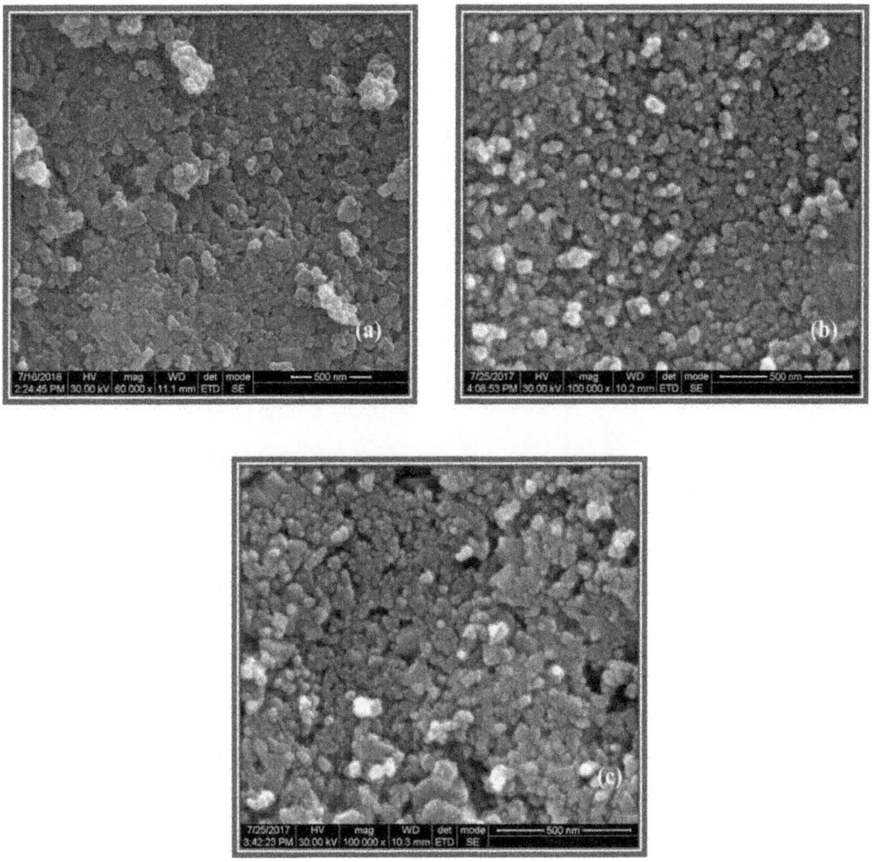

Fig. 4.3: HRSEM images for (a) MnO_2 (b) α-Fe_2O_3 and (c) MnO_2/α-Fe_2O_3 Nanocomposite.

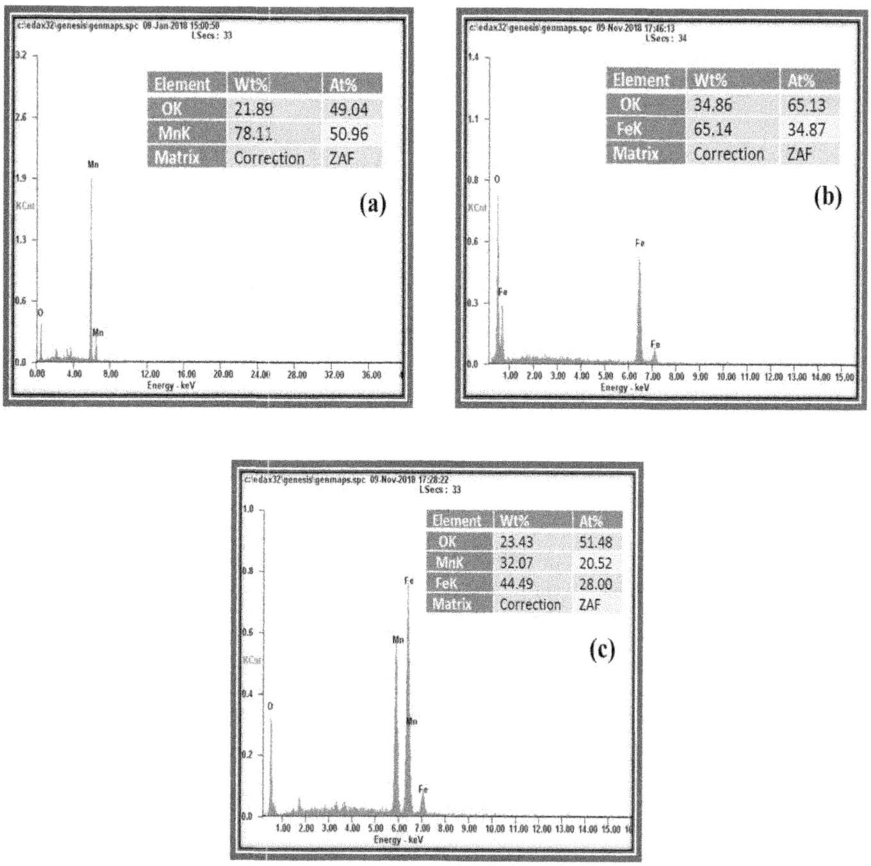

Fig. 4.4: EDX Spectrum of (a) MnO_2 (b) α-Fe_2O_3 and (c) MnO_2/α-Fe_2O_3 Nanocomposite.

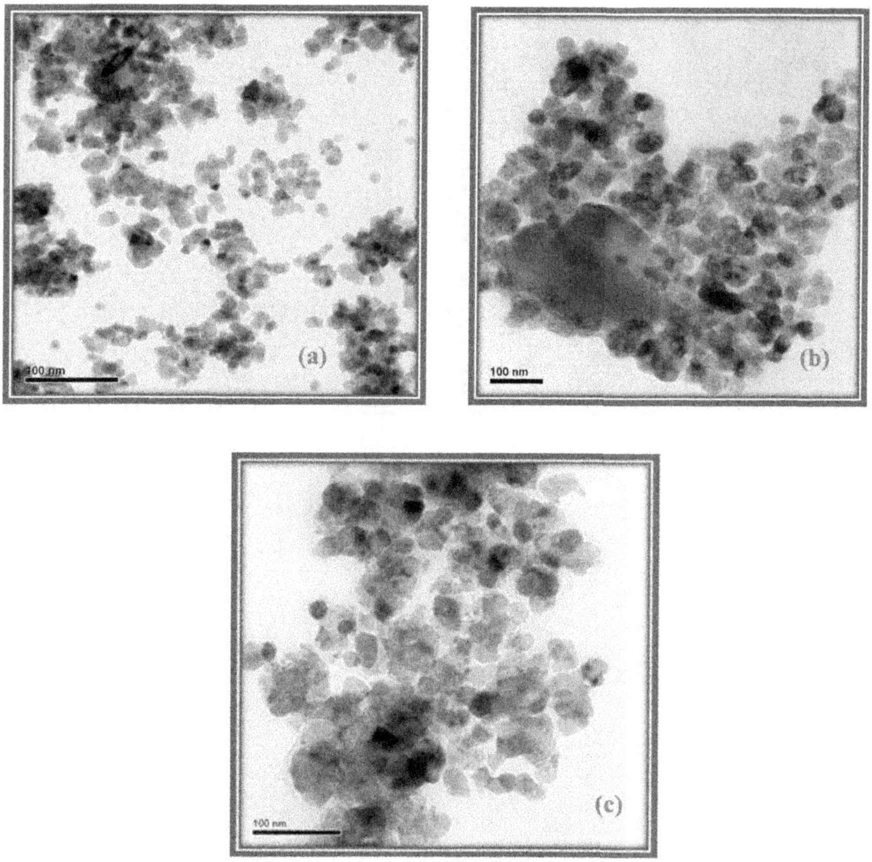

Fig. 4.5: TEM Images of (a) MnO_2 (b) α-Fe_2O_3, and (c) MnO_2/α-Fe_2O_3 Nanocomposite.

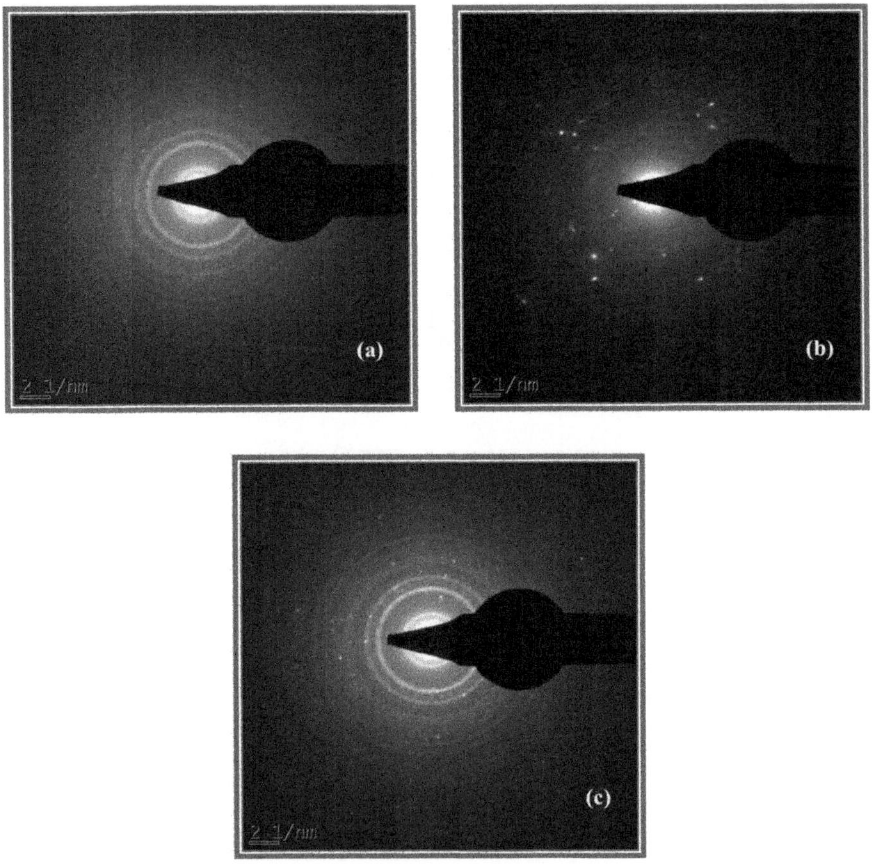

Fig. 4.6: SAED Pattern of (a) MnO_2 (b) α-Fe_2O_3 and (c) MnO_2/α-Fe_2O_3 Nanocomposite.

This can also be accredited to the surface effects includes spin canting, decrease in Fe content, partial oxidation and deviations from stoichiometry [Darezereshki. E. 2011, Lassoued. A et al. 2017]. The magnetic hysteresis loop of MnO_2 is presented in Fig. 4.7(a), the saturation magnetizations (Ms) and coercivity (Hc) of MnO_2 nanoparticles was found to be 0.830 emu/g and 62.73 Oe respectively. At low external field, the MnO_2 nanoparticles reveals high coercivity ($Hc = 62.73$ Oe), which is characteristic behaviour of weak ferromagnetism. At high external field, magnetization dependence on applied field is significantly linear and Ms is not reached even at the maximum applied field of 500 Oe. From the perspective view of magnetic interactions, only Mn^{4+}–O^{2-}–Mn^{4+} is in ferromagnetic coupling, and other magnetic interactions are correspondingly in antiferromagnetic coupling behaviour. Therefore, the magnetic response of nanostructured MnO_2 could be ascribed to the ferromagnetic coupling of Mn^{4+}–O^{2-}–Mn^{4+}, because the magnetic properties are determined by magnetic interactions between the Mn ions, which in turn rest on the Mn valence ion distribution in the $[Mn_2]O_4$ framework [Masquelier. C et al. 1996]. Fig. 4.7(c) presents the magnetic hysteresis loop of MnO_2/α-Fe_2O_3 nanocomposites. The saturation magnetizations (Ms) and coercivity (Hc) of composites was found to be 0.896 emu/g and 1255.36 Oe respectively.

4.3.5 X-ray Photoelectron Spectra (XPS) Analysis

The elemental composition, electronic and chemical states of MnO_2/α-Fe_2O_3 nanocomposite materials were determined using X-ray photoelectron spectroscopy (XPS) analysis. Fig. 4.8(a) shows the full survey spectrum of MnO_2/α-Fe_2O_3 nanocomposite and Fig. 4.8(b-e) displays the individual scanned spectra of C 1s, Mn 2p, Fe 2p and O 1s elements in the composite sample. Fig. 4.8(a) depicts the wide-scan XPS survey spectrum of nanocomposites makes evident the existence of Mn (Manganese), Fe (Iron), and O (Oxygen) elements in the nanocomposites. The survey spectrum from Fig. 4.8(c) shows the existence of Mn and O, with C from the reference. Moreover, the atom molar rate of Mn and O is about 1:2 in the survey spectra, demonstrating the formation of MnO_2. There are two XPS binding energies of Mn $2p_{1/2}$ and Mn $2p_{3/2}$ at 652.72 eV and 641.20 eV were observed for MnO_X.

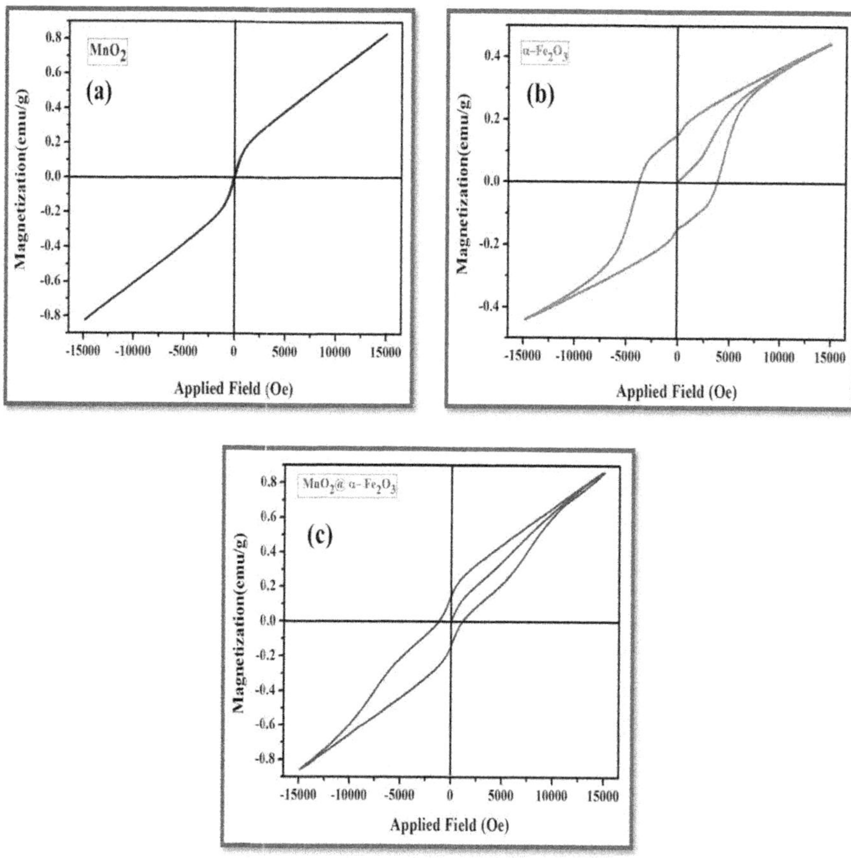

Fig. 4.7: Magnetic Hysteresis Curves of (a) MnO_2 (b) α-Fe_2O_3 and (c) MnO_2/α-Fe_2O_3 Nanocomposite.

It can be seen that the separation of binding energy is 11.5 eV among the Mn $2p_{1/2}$ and Mn $2p_{3/2}$ states. Moreover, Mn $2p_{1/2}$ can be separated into two characteristic peaks which can be ascribed to Mn^{4+} and Mn^{3+} ions and Mn $2p_{3/2}$ can be also separated into two characteristic peaks which could be attributed to Mn^{4+} and Mn^{3+}. The higher oxidation state manganese species are more active for the electrochemical performance over manganese-based electrodes. In Fig. 4.8(d), two prominent peaks at 708.95 and 722.56 eV were observed corresponding to Fe $2p_{3/2}$ and Fe $2p_{1/2}$ respectively. A satellite peak was detected in gap between two prominent peaks, which is confirming that the prepared iron oxide nanoparticles are belonged to α-Fe_2O_3 as opposed to Fe_3O_4 [Wang. R et al. 2014]. The O1s spectra of the MnO_2/α-Fe_2O_3 nanocomposites in Fig. 4.8(e) were deconvoluted using a curve-fitting technique, and contain two kinds of distinguished surface oxygen species. The lower binding energy peak at 528.0–530.0 eV might be consigned to the surface lattice oxygen (O_β), such as defective oxides or surface oxygen ions bonded to manganese in a coordinately saturated environment, and the higher binding energy peak at 530.0 to 532.0 eV could belong to the surface chemisorbed oxygen (Oα) [Tang. X.F et al. 2010] Generally, the surface chemisorbed oxygen (Oα) is more reactive than the surface lattice oxygen (O_β), which is due to the higher mobility of Oα. In Fig. 4.8(e), the peak at 528.52 eV could be assigned to O^{2-} and the peak at a higher binding energy (530.73 eV) might be endorsed to surface hydroxyl groups [Wang. R et al. 2014].

4.3.6 Electrochemical Studies

Electrochemical properties of MnO_2/α-Fe_2O_3 nanocomposite were evaluated using a three-electrode system in 1M aqueous KOH electrolyte with the reference electrode (Ag/AgCl) and a counter electrode (platinum foil). For comparison, electrochemical behavior of pristine MnO_2 and α-Fe_2O_3 were also investigated. Fig. 4.9 shows the CV curves of the pristine MnO_2, α-Fe_2O_3 and MnO_2/α-Fe_2O_3 nanocomposite. The CV measurement of an electrode was taken in potential range of -0.1 V to 0.4 V. The two regions in CV plot indicate the area of negative current reduction at the cathode and positive current region of anode oxidation.

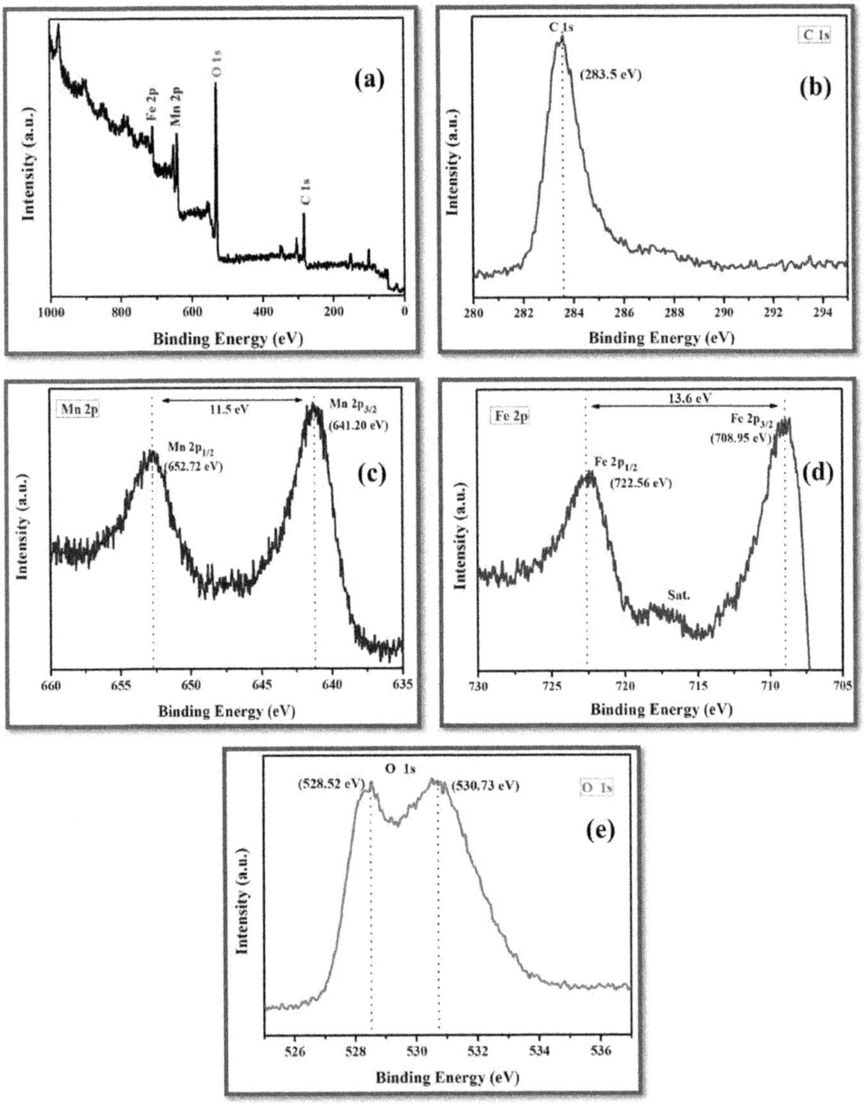

Fig. 4.8: XPS spectra of MnO_2 /α-Fe_2O_3 nanocomposites (a) Full survey scan spectrum and (b–e) individual XPS spectra of C1s, Fe 2p, Mn 2p and O 1s respectively.

Fig. 4.9(b) illustrates the CV diagrams of pristine α-Fe_2O_3 active electrode. All CV curves showed a pair of redox peaks at various scan rates from 5–320 mV/s over the voltage range of -0.1–0.4 V, indicating similar pseudocapacitive response and significant rate capability. Additionally, when increasing of scan rates, the redox peaks current intensity areas turn into larger as the scan sweep has increased [Sarkar. D et al. 2017, Zhong. Y et al. 2016, Zhong. J et al. 2017]. Moreover, enormously shifting in anodic peaks towards a positive (+ve) potential owing to electroactive sample polarization, while the cathodic peaks moved towards a negative (–ve) potential with increasing scan rate since the low resistance and fast ion and electron transfer rates. The strong peaks can be closely correlated to the morphology and chemical composition of α-Fe_2O_3 electrode material for the reversible faradaic redox reactions and redox reaction resultant to KOH electrolyte. Fig. 4.9(a) depicts the pristine MnO_2 CV curve, which appears as rectangular shape without obvious redox peaks; capacitive behavior of MnO_2 nanospheres electrode is pseudocapacitance, attributed to reversible redox reaction among Mn^{3+} and Mn^{+4}. The plot shape is also not perfectly rectangular, which point out a deviation from ideal behavior of the supercapacitor [Liu. Li et al. 2016]. The slight shape change in CV curves specifies the good charging rate capability of the electrode. Fig. 4.9(c) appeared CV curves of pristine MnO_2/α-Fe_2O_3 electrode materials with various mass loadings of MnO_2/α-Fe_2O_3 nanoparticles at various scan rates from 5–320 mV/s. It could be realized that the CV curves of pristine MnO_2/α-Fe_2O_3 electrode show relatively apparent redox peaks, which can be ascribed to typical pseudocapacitive performance of MnO_2/α-Fe_2O_3 electrode. As the gravimetric current density increases with the loading level of MnO_2/α-Fe_2O_3 nanoparticles, coupled with the apparent redox peaks, it is clear that MnO_2/α-Fe_2O_3 nanoparticles in fact show the part of pseudocapacitors [Jia. R et al. 2017]. The specific capacitance performance of the as-synthesized sample was also investigated by using the charge-discharge method and the results of pristine MnO_2, α-Fe_2O_3 and MnO_2/α-Fe_2O_3 nanocomposite are shown in Fig. 4.10. The galvanostatic charge–discharge (GCD) test was performed at a current density of 1 Ag^{-1}. The voltage range was maintained from -0.1 to 0.4 V during the GCD test.

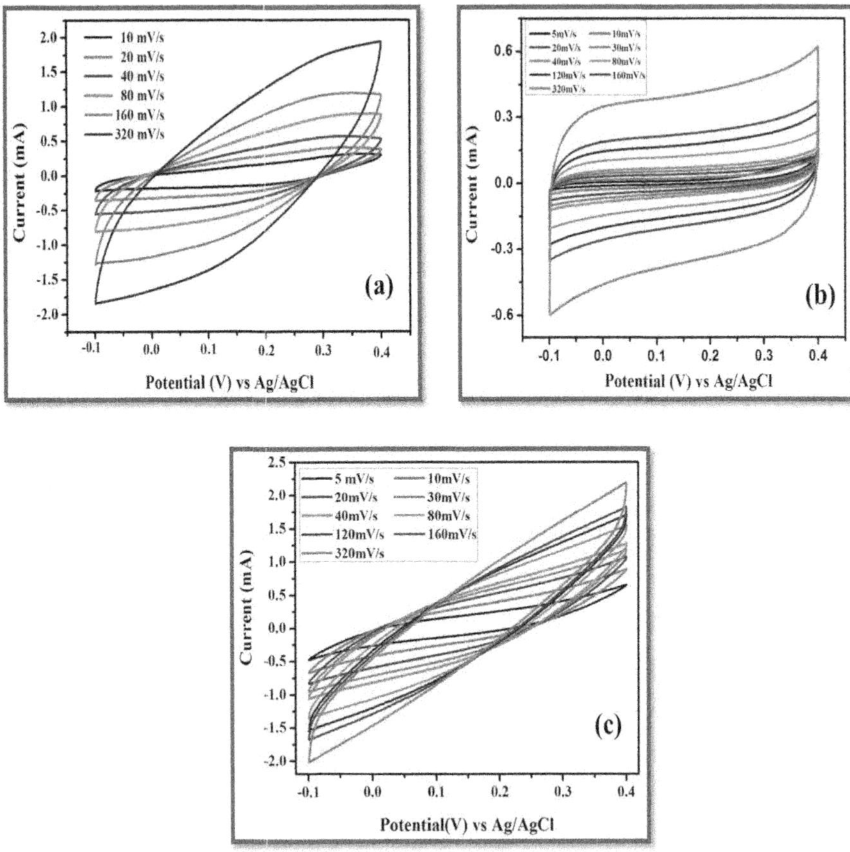

Fig. 4.9: Cyclic Voltammograms plots at different scan rates of (a) MnO_2 (b) α-Fe_2O_3, and (c) MnO_2/α-Fe_2O_3 nanocomposite.

The specific capacitance from the GCD plot can be calculated by the equation:

$$\textbf{Specific Capacitance} = C_{GCD} = \frac{(i\times\Delta t)}{(\Delta v\times m)} \quad \textbf{(4.1)}$$

Where *i/m* is the current density (Ag^{-1}), Δt corresponds to discharge time and Δv stand for potential range. The charge/discharge curves of $MnO_2/\alpha\text{-}Fe_2O_3$ electrode displays that the curve exhibits equilateral triangle shapes which indicate the good reversibility during the charge/discharge process [Yang. S.H et al. 2014]. The calculated specific capacitances of $MnO_2/\alpha\text{-}Fe_2O_3$ are 216.35, 128.47, 77.80 and 60.93 Fg^{-1} at current densities of 4, 2, 1 and 0.5 A/g, respectively (Fig. 4.10(c)). And those of $\alpha\text{-}Fe_2O_3$ are 116.99, 58.61, 38.30 and 29.72 Fg^{-1} at current densities of 4, 2, 1 and 0.5 A/g respectively (Fig. 4.10(b)). The corresponding specific capacitances of MnO_2 at current densities of 4, 2, 1 and 0.5 A/g were calculated as 37.47, 45.15, 66.78, and 175.51 F/g respectively (Fig. 4.10(a)) which are more than the reported values in earlier reports.

This result declares the specific capacitance of $MnO_2/\alpha\text{-}Fe_2O_3$ electrode is much greater than pristine MnO_2 and $\alpha\text{-}Fe_2O_3$ electrode nanoparticles. The higher specific capacitance of $MnO_2/\alpha\text{-}Fe_2O_3$ electrode is due to MnO_2 spherical nanoparticles were well wrapped on the surface of $\alpha\text{-}Fe_2O_3$ nanoparticles to provide composites with larger surface area [Wang. D et al. 2011]. Our material also showed a better specific capacitance than the specific capacitance of MnO_2 electrode based composites. The energy density and power density of the composites were calculated using Equation 4.2 and 4.3 and tabulated in the Table 4.1.

$$\mathbf{E_d = 0.5C_s\Delta V^2} \quad \textbf{(4.2)}$$

$$\mathbf{P_d = \frac{E_d}{t}} \quad \textbf{(4.3)}$$

Where, E_d is energy density (Wh kg^{-1}), C_s is specific capacitance (Fg^{-1}), ΔV is potential window of discharge (V), P_d is power density (W kg^{-1}) and Δt is discharge time (s). The energy density is found to increase with the power density. The power density is in increasing trend which point out the pseudocapacitor discharging capacity is directly proportional to discharge current.

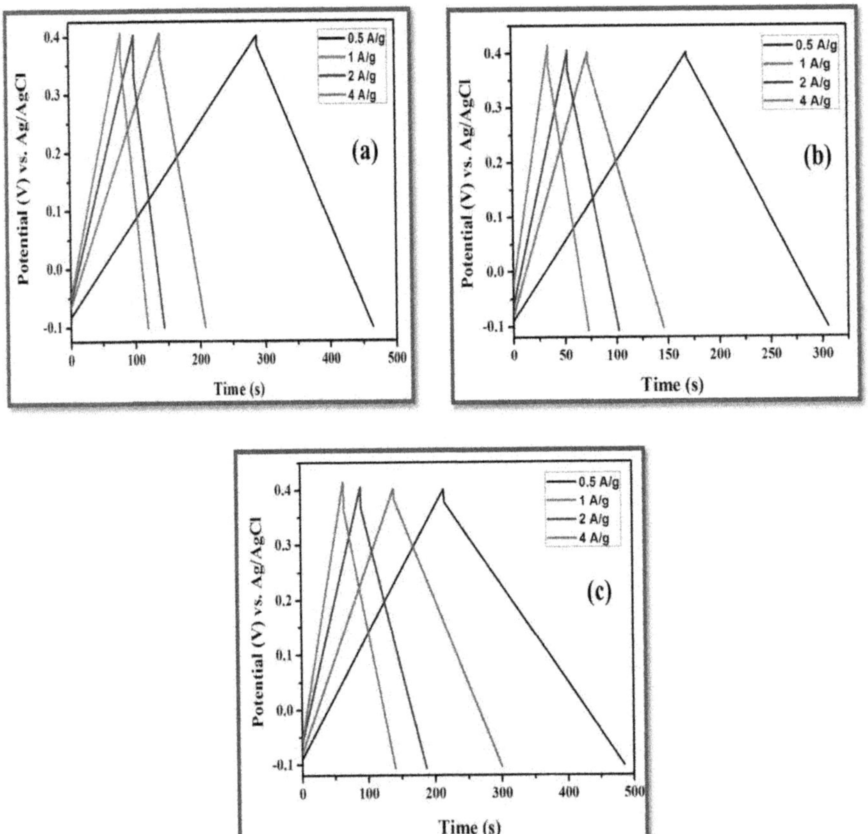

Fig. 4.10: Charge–Discharge Curves of (a) MnO_2 (b) α-Fe_2O_3 and (c) MnO_2/α-Fe_2O_3 nanocomposite at different current densities.

Table 4.1: Specific Capacitance, Energy and Power density values of MnO_2, α-Fe_2O_3, and MnO_2/α-Fe_2O_3 Nanocomposite

Sample	Current density (Ag^{-1})	Specific Capacitance (Fg^{-1})	Energy density ($Wh \cdot kg^{-1}$)	Power density ($kW \cdot kg^{-1}$)
MnO_2	0.5	175.51	688.0270	0.1508
	1	66.78	261.7838	0.3016
	2	45.15	176.6254	0.6031
	4	37.47	146.8871	1.2062
α-Fe_2O_3	0.5	116.99	458.6317	0.2083
	1	58.61	229.7574	0.4166
	2	38.30	150.1391	0.8333
	4	29.72	116.5133	1.6665
MnO_2/α-Fe_2O_3	0.5	216.35	848.092	0.2869
	1	128.47	503.6274	0.5737
	2	77.80	304.9807	1.1474
	4	60.93	238.8816	2.2949

For pure α-Fe_2O_3 and MnO_2 energy density of 458.6317 Wh kg^{-1} and power density of 0.2083 kW kg^{-1} and energy density of 688.0270 Wh kg^{-1} and power density of 0.1508 kW kg^{-1} were observed at a current density of 0.5 Ag^{-1}. At the same current density of 0.5 Ag^{-1}, the MnO_2/α-Fe_2O_3 nanocomposites provided a superior energy density of 848.092 Wh kg^{-1} and power density of 0.2869 kW kg^{-1}. The calculated measurements of specific capacitance, energy and power density values of pure MnO_2, α-Fe_2O_3 and MnO_2/α-Fe_2O_3 nanocomposites are tabularized in Table 4.1. Table 4.2 summaries the comparisons of specific capacitance performance of MnO_2/α-Fe_2O_3 nanocomposites prepared in the present work revealed good cycling performance than some of the previously reported papers [Cho. S et al. 2018]. From the GCD plot, the Coulombic efficiency was calculated by following equation:

Coulombic Efficiency = [discharging time (t_D)/charging time (t_C)] ×100%. (4.4)

Coulombic efficiency (calculated with respect to the second GCD cycle) of MnO_2/α-Fe_2O_3 nanocomposites which is good for potential supercapacitor applications. As exposed in Fig. 4.12, the MnO_2/α-Fe_2O_3 nanosphere-like structure exhibited excellent long-term electrochemical stability and high rate stability with a very slight decrease to 85.4%, even after 10000 cycles. Such exceptional cycling stability of MnO_2/α-Fe_2O_3 electrode may be endorsed to the following: the sphere like structure, which can improve the volume changes during the charge–discharge process to promise good stability; gradual penetration of the KOH electrolyte ions into an electro-active material; and reduced volume expansion resulting from the rapid and long-term faradaic reaction. All of the results indicate that the MnO_2/ α-Fe_2O_3 nanocomposite is a suitable electrode material for supercapacitors.

Specific capacitance vs increasing current density was conspired in Fig. 4.11. It is clearly concluded that specific capacitance decreases with an increase in current density due to the fact that high current density prevents the accessibility of ions from entering into the inner core of the electrode material. At a low current density, the transport of ions is restricted due to slow diffusion and only ions at outer surfaces could be used for charge storage. The obtained specific capacitance values for the MnO_2/α-Fe_2O_3 nanocomposite electrode are 216.35, 128.47, 77.80, and 60.93 F/g for different current density values of 0.5, 1, 2 and 4 A/g respectively.

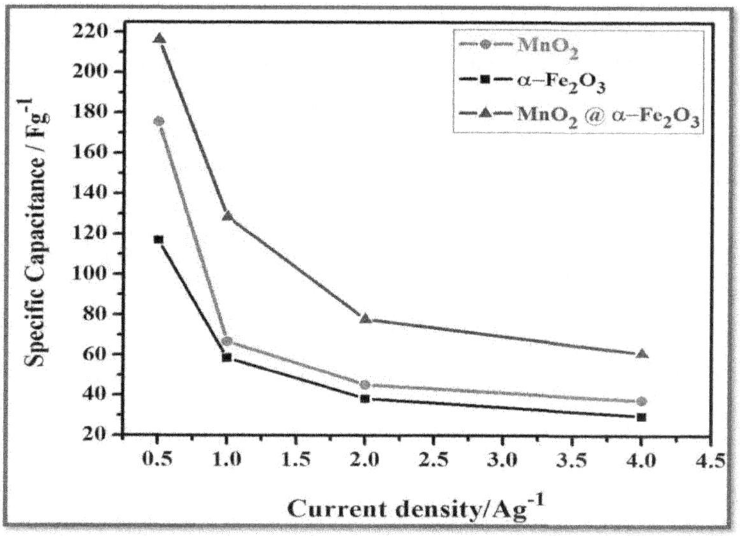

Fig. 4.11: Specific Capacitance as a Function of Current density of Pure of MnO_2, α-Fe_2O_3 and MnO_2/α-Fe_2O_3 Nanocomposite Electrodes using GCD Technique.

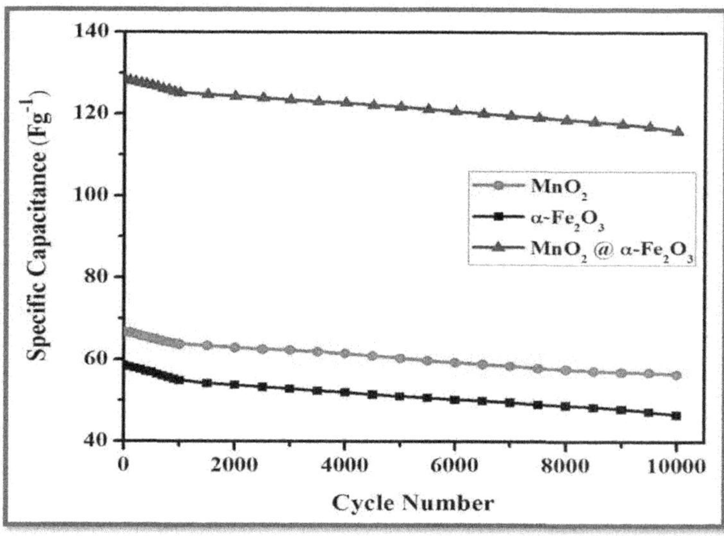

Fig. 4.12: Cycling Performance of the MnO_2, α-Fe_2O_3 and MnO_2/α-Fe_2O_3 nanocomposite electrodes.

Table 4.2: Comparison of maximum capacitance values of the reported and the present electrode of $MnO_2/\alpha\text{-}Fe_2O_3$ nanocomposite

Electrode Material	**Morphology**	**Electrolyte**	**Capacitance (Fg^{-1})**	**Ref. No**
$\alpha-Fe_2O_3/MnO_2$ core shell	Nanorods	Na_2SO_4 (1M)	227mFg^{-1} (0.5mA/cm^2)	[Darezereshki. E. 2011]
$\alpha-Fe_2O_3$@MnO_2	Nanotubes	KOH (3M)	289.9 Fg^{-1} (1Ag^{-1})	[Kai. K et al. 2012]
MnO_2@Fe_2O_3 nanospindles	Nanospindle	K_2SO_4 (0.5M)	159 Fg^{-1} (0.1Ag^{-1})	[Sarkar. D et al. 2013]
FeS_2@Fe_2O_3 hybrids	Nanospheres	Li_2SO_4 (1 M)	255Fg^{-1} (1 Ag^{-1})	[Lassoued. A et al. 2017]
MnO_2/graphene	Honeycomb	KOH (1M)	278.5 Fg^{-1} (1Ag^{-1})	[Masquelier. C et al. 1996]
Fe_2O_3@N-doped graphene	Nanoparticle	KOH (2M)	274 Fg^{-1} (1 Ag^{-1})	[Sarkar. D et al. 2017]
$NiCO_2S_4$@Fe_2O_3	Nanorod	Na_2SO_4 (1M)	285 Fg^{-1} (2.5Ag^{-1})	[Zhong. Y et al. 2016]
$\alpha-Fe_2O_3$ /G	Spherical	Na_2SO_4 (1M)	306.9 Fg^{-1} (3Ag^{-1})	[Zhong. J et al. 2017]
$\alpha-Fe_2O_3$	Nanosheets	Li_2SO_4(1 M)	116 Fg^{-1} (0.75 Ag^{-1})	[Liu. Li et al. 2016]
MnO_2 and $\alpha\text{-}Fe_2O_3$	Nanoflakes	$LiClO_4$ (0.5M)	261 Fg^{-1} (1Ag^{-1})	[Jia. R et al. 2017]
$MnO_2/\alpha\text{-}Fe_2O_3$ nanocomposites	**Spherical**	**KOH (1M)**	**216.35 (0.5 Ag^{-1})**	**This Work**

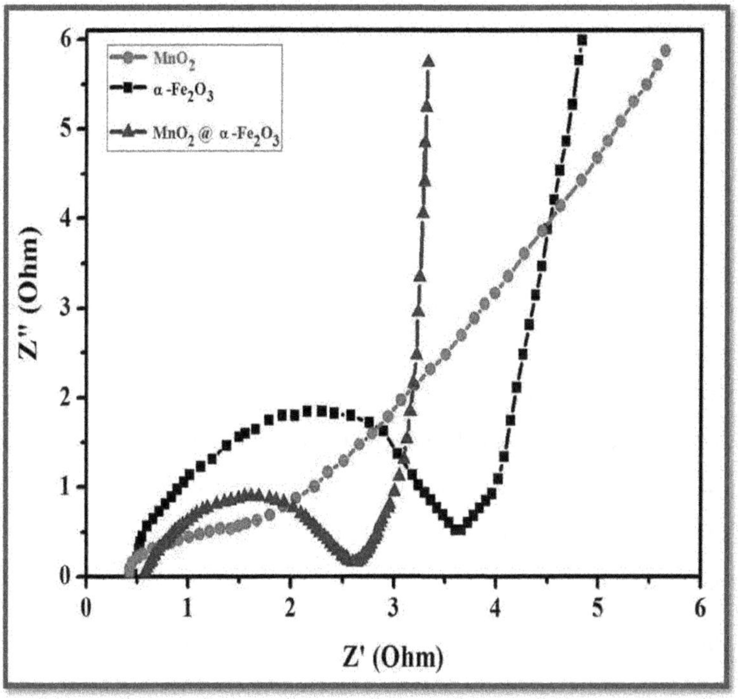

Fig. 4.13: Nyquist Plots of MnO_2, α-Fe_2O_3 and MnO_2/α-Fe_2O_3 nanocomposite electrodes.

These results indicate that $MnO_2/\alpha\text{-}Fe_2O_3$ nanocomposite electrode has excellent specific capacitance at lower current densities and higher energy density and power densities at higher current density of 4 Ag^{-1}.

Electrochemical impedance spectroscopy (EIS) can prove the transmission ability of ions and charges in supercapacitors. The electrochemical actions of the MnO_2, $\alpha\text{-}Fe_2O_3$ and $MnO_2/\alpha\text{-}Fe_2O_3$ nanocomposite electrodes were studied using EIS analysis. The Nyquist impedance plot for the nanocomposite electrode is displayed in Fig. 4.13. The Nyquist impedance plot for the prepared nanocomposite electrode contains of a semicircle at the higher frequency region and a linear line at lower frequency region. The semicircle at the higher frequency region could be characteristics of the resistance of oxide-electrolyte interface. The resistance contribution from the faradaic redox procedure is involved in this region. The lower frequency region can be the characteristics of Warburg impedance, which defines diffusive resistance of the OH- ions inside the electrode material. The plot displays the linear line along the y-axis, which is attributed to the Faradaic behaviour owing to the reversible Faradaic redox reaction. All of the results point out that the $MnO_2/\alpha\text{-}Fe_2O_3$ nanocomposite is a suitable electrode candidate material for supercapacitors.

4.4 CONCLUSION

In summary, a simple and easy presentable hydrothermal route was implemented to fabricate $MnO_2/\alpha\text{-}Fe_2O_3$ nanocomposite. The positive synergism between MnO_2 and $\alpha\text{-}Fe_2O_3$ nanoparticles leads to the formation of mechanically robust, electrically conductive and thermally stable composite. Further, the composite was applied as a working electrode material to demonstrate its potential candidature to be a promising electrode substance for high performance hybrid supercapacitors. The synthesized $MnO_2/\alpha\text{-}Fe_2O_3$ nanocomposite was investigated using XRD, HRSEM, EDX, TEM, VSM and XPS. The FTIR spectra of MnO_2, $\alpha\text{-}Fe_2O_3$ and its nanocomposite confirm the formation of $MnO_2/\alpha\text{-}Fe_2O_3$ nanocomposite and interaction between MnO_2 and $\alpha\text{-}Fe_2O_3$ nanocrystalline. The electrochemical achievements of prepared electrodes were studied through CV and galvanostatic charge/discharge (GCD) test. The CP curves of nanocomposites shows the linear nature of discharge curve and substantially low IR drop clearly indicate a good capacitive behavior of the nanocomposite. The $MnO_2/\alpha\text{-}Fe_2O_3$ nanocrystalline composite exhibits a high specific capacity of 216 Fg^{-1} at a current density of 0.5 Ag^{-1}, energy density of 848.092 $Wh\cdot kg^{-1}$and power density value measured of 0.2869 kW kg^{-1} compound with good rate capacity and cyclic stability. The better specific capacitance of $MnO_2/\alpha\text{-}Fe_2O_3$ nanocomposites compared to pure MnO_2, $\alpha\text{-}Fe_2O_3$,may be due to the presence of, MnO_2 having high surface area in the nanocomposites, good interaction between $\alpha\text{-}Fe_2O_3$ and MnO_2 which gives a superior structure easy electrolyte accessibility. In addition, these results are supported by the EIS analysis, which suggests high phase homogeneity, high capacitance and improved charge-transfer characteristics. All the above results imply that the MnO_2, $\alpha\text{-}Fe_2O_3$, and $MnO_2/\alpha\text{-}Fe_2O_3$ nanocomposite materials can be good inexpensive electrode materials for next generation supercapacitor applications.

CHAPTER V

SUMMARY AND FUTURE WORK

5.1 SUMMARY

With the rapid development of social economy and the continuous advancement of industrialization, people's demand for efficient energy storage has become more and more urgent, which leads to more and more researchers engaged in the study of energy storage devices based on electrochemical properties. The supercapacitor is one of the most promising sustainable energy storage devices of the century, which have attracted great attention because of its fast charge and discharge rates, high power density, good reversibility and long cycling life. These features make supercapacitor promising as a potential energy storage device, and it is used in new energy vehicles, solar energy systems and smart grid. The electrode material is an important factor to determine the performance of supercapacitors. At present, electrode materials are roughly classified into four categories: transition metal oxides, carbon materials, conductive polymers, metal sulphides, and others. Compared with the other materials, transition metal oxides are favoured by researchers because of their low cost and extremely high theoretical specific capacitance. Transition metal oxides such as RuO_2, MnO_2, NiO, V_2O_5, CO_3O_4, and TiO_2 are widely applied in pseudocapacitive supercapacitor due to its high specific capacitance, energy density and power density. Among these materials, MnO_2 is a promising transition metal oxide materials as a supercapacitor electrode because of electrochemical stability, non-toxic, natural abundance, low cost and pseudocapacitive. However, its specific capacitance decreases seriously with the increase of scanning rate because of the characteristics of poor conductivity. Therefore, the combinations of MnO_2 with other transition metal oxide materials are required to effectively improve the properties of the materials due to synergistic effect between the oxides.

MnO_2 based composite materials have many advantages, which makes them promising materials for advanced SCs. It seems that SC electrode materials based on MnO_2 will occupy more and more prominent position in the research of SCs.

We presented the electrochemical investigation, discussion and future work related to various kinds of manganese dioxide based composite materials, such as manganese dioxide /nickel oxide, manganese dioxide/ copper oxide and manganese dioxide /Iron oxide composites.

The first chapter comprehensively highlighted and discussed the basic mechanism, types, characterization techniques of supercapacitors, the most important advances in metal oxide and more specifically MnO_2–based aqueous supercapacitors over the recent years, spanning from fundamental charge storage mechanisms, electrode materials engineering and functionalization, to tailored supercapacitor devices for multipurpose applications. Significant achievements have been attained during that last few years, including deeper understanding of the mechanisms governing charge storage, different electrode engineering routes to enhance electrochemical performance, and the design of flexible and/or smart devices.

The first nanocomposite namely MnO_2/NiO was successfully prepared by simple hydrothermal method to investigate the electrochemical performance of supercapacitor. The synthesized MnO_2/NiO nanocomposite was characterized using XRD, HRSEM, EDX, TEM, VSM and XPS analysis. The electrochemical achievements of prepared electrodes have been studied through cyclic voltammetry (CV) and galvanostatic charge/discharge (GCD) test. The MnO_2/NiO nanocomposite reveals a stable cycle performance and notable rate performance. The structural and surface morphology studies of MnO_2/NiO nanocomposite have revealed nanocrystalline nature. The increased conductivity and uniform morphology of the compound accelerates transfer of electrons and the electrolyte diffusion in the hybrid compounds. The MnO_2/NiO nanocrystalline composite shows a high specific capacity of 247 Fg^{-1} at a current density of 0.5 Ag^{-1}, energy density of 971.2948 $Wh{\cdot}kg^{-1}$and power density of 0.1346 kW kg^{-1} compound with good rate capacity and cyclic stability. Enhanced electrochemical performance of MnO_2/NiO than pristine MnO_2 and NiO could be considered as potential electrode materials for next generation supercapacitors.

MnO_2/CuO nanocomposite was successfully prepared via a simple hydrothermal method. The morphology, structure, composition and electrochemical behavior of the as prepared materials were studied using XRD, HRSEM, EDX, TEM and XPS analysis. The electrochemical performance of the MnO_2/CuO nanocomposite system has been investigated at different scan rates in (CV) and current densities (GCD) with a three-electrode system in 1M aqueous KOH electrolyte with reference electrode (Ag/AgCl) and counter electrode (Pt foil). Firstly, CV measurements were performed for various scan rates (5 to 320 mV/s) within a potential window of 0.1 to 0.4 V. All CV curves showed a pair of redox peaks at various scan rates, indicating pseudocapacitive response and significant rate capability. The resultant MnO_2/CuO nanocomposite exhibited the specific capacitance values are 279.12, 161.57, 89.48 and 66.21 F/g at current densities of 0.5, 1, 2 and 4 A/g, respectively and retaining 91.26% of the *Cs* value of after 10000 cycles thus showing outstanding cycling stability. The measured specific capacitance values of MnO_2/CuO (279.12 F/g) electrode are much greater than their individual compound of bare MnO_2 (175.51 F/g) and CuO (191.06 F/g) electrode materials. The improvement in specific capacitance of MnO_2/CuO electrode is due to CuO spherical nanoparticles were well wrapped on the surface of MnO_2 nanoparticles to provide composites with larger surface area.

A simple hydrothermal route was implemented to fabricate $MnO_2/\alpha\text{-}Fe_2O_3$ nanocomposite as supercapacitor electrode material. The synthesized $MnO_2/\alpha\text{-}Fe_2O_3$ nanocomposite was investigated using XRD, HRSEM, EDX, TEM, VSM and XPS. The electrochemical studies were inspected by CV, GCD and EIS techniques. The results emphasize that the $MnO_2/\alpha\text{-}Fe_2O_3$ nanocomposites electrodes exhibit high rate capability, high energy density, high power density and also long cycle stability, which make this type of environment friendly, nanostructures a perfectly promising material for fabrication of supercapacitors electrode. The CV measurement of an electrode was taken in potential range of -0.1 V to 0.4 V. All CV curves showed a pair of redox peaks at various scan rates from 5–320 mV^{-1} over the voltage range of-0.1–0.4 V, indicating similar pseudocapacitive response and significant rate capability. The GCD test was performed at a increasing current densities of 0.5, 1, 2 and 4 A/g and maintained -0.1to 0.4 V range during entire GCD test. The MnO_2/α-Fe_2O_3 nanocrystalline composite exhibits a high specific capacitance are found to be 216.35, 128.47, 77.80 and 60.93 F g^{-1} at current densities of 4, 2, 1 and 0.5 A/g, respectively. The high specific capacitance of MnO_2/α-Fe_2O_3 nanocomposites compared to pure α-Fe_2O_3, MnO_2 system may be due to the presence of, MnO_2 having high surface area in the nanocomposites, good interaction between α-Fe_2O_3 and MnO_2 which gives a superior structure easy electrolyte accessibility. All the above results imply that the MnO_2/α-Fe_2O_3 nanocomposite materials can be good inexpensive electrode materials for next generation supercapacitor applications.

The present investigation provides new insight to fabricate metal oxide based pseudocapacitance behavioural electrode materials for high performance energy storage devices. An excellent electrochemical performance of the MnO_2 based composite electrode is attributed to the synergy realized between the pseudocapacitive charge storage mechanism and the involvement of redox couple catalysed electrochemical reactions at the electrode surface. This study establishes the MnO_2/NiO, MnO_2/CuO and MnO_2/α-Fe_2O_3 composite system as a better supercapacitor electrode design than many other MnO_2 based supercapacitor electrodes reported earlier.

5.2 SUGGESTIONS FOR FUTURE WORK

- The present study was focused on the synthesis of metal oxide based nanocomposites. In future, attempts can be made to synthesize different morphologies and their impact can be investigated. The synthesis route developed in the present work for the fabrication of the metal composites is cost effective and environmental friendly hydrothermal method, which could be extended to prepare different metal oxides composites.

- The electrochemical behaviours of the nanocomposites were investigated using aqueous electrolyte. In future, attempts can be made to evaluate their performance in various non-aqueous and polymer electrolytes.

- In future, attempts can be made to incorporate other carbon materials like single walled carbon nanotubes (SWCNTs), multi walled carbon nanotubes (MWCNTs) and activated carbon for enhancing the specific capacitor values.
- In the present work, the electrochemical performances of the developed electrode materials were evaluated in three-electrode system. In future, attempts can be made to fabricate symmetric and asymmetric supercapacitors using the different electrodes. Further, attempts can be made to employ the developed nanocomposites in lithium ion batteries and solar cell applications.
- With the continuous efforts on the research of metal oxide/polymer nanocomposites for supercapacitors, it is anticipated that the above discussed issues could be solved properly in the coming years.
- New types of supercapacitor devices such as microsupercapacitors and fiber-type supercapacitors have been widely demonstrated, which can meet the energy requirements of the next generation of smart electronic devices. Pushing further the size limit of micro-supercapacitors is needed towards more advanced on chip microelectronic devices. 3D printing could be very a useful technique for the fabrication of microsupercapacitors.
- Nowadays, developing free-standing, flexible supercapacitors that can remain excellent electrochemical performance under external mechanical deformation such as bending, twisting, stretching and compression, and/or against harsh environments such as very low/high temperatures or stress, is one of the hottest directions in flexible electronics. Safe operation should be always taken into seriously account especially for applications in wearable electronic devices and smart textiles. Integrated multifunctional systems combining supercapacitor devices with other kinds of energy harvesting or energy storage devices, such as solar cells, batteries, Nano generators, or electrochromic devices to realize energy conversion and storage in a single unit, is an emerging direction to fabricated advanced smart energy devices.

9 781805 250746

Printed by BoD™ in Norderstedt, Germany